The Minerals, Metals & Materials Series

Jinichiro Nakano · P. Chris Pistorius
Candan Tamerler · Hideyuki Yasuda
Zuotai Zhang · Neslihan Dogan
Wanlin Wang · Noritaka Saito
Bryan Webler
Editors

Advanced Real Time Imaging II

Cutting-Edge Techniques in Materials Science Studies

TMS

Springer

Editors
Jinichiro Nakano
National Energy Technology Laboratory/Leidos
Albany, OR, USA

P. Chris Pistorius
Carnegie Mellon University
Pittsburgh, PA, USA

Candan Tamerler
The University of Kansas
Lawrence, KS, USA

Hideyuki Yasuda
Kyoto University
Kyoto, Japan

Zuotai Zhang
South University of Science and Technology
Shenzhen, China

Neslihan Dogan
McMaster University
Hamilton, ON, Canada

Wanlin Wang
Central South University
Changsha, Hunan, China

Noritaka Saito
Kyushu University
Fukuoka, Japan

Bryan Webler
Carnegie Mellon University
Pittsburgh, PA, USA

ISSN 2367-1181 ISSN 2367-1696 (electronic)
The Minerals, Metals & Materials Series
ISBN 978-3-030-06142-5 ISBN 978-3-030-06143-2 (eBook)
https://doi.org/10.1007/978-3-030-06143-2

Library of Congress Control Number: 2018964684

Cover image courtesy of Anna Nakano, National Energy Technology Laboratory

This Springer imprint is published by the registered company Springer Nature Switzerland AG
The registered company address is: Gewerbestrasse 11, 6330 Cham, Switzerland

Contents

About the Editors

Jinichiro Nakano is Principal Research Scientist working for the US Department of Energy National Energy Technology Laboratory. His expertise includes in-situ high-temperature experimental investigations and computational thermodynamics in a broad range of materials with an emphasis on energy and environmental applications. His publications have been repeatedly ranked in the Elsevier's ScienceDirect TOP25 Hottest Articles and Most Cited/Downloaded Articles. His work has been internationally recognized by the Gustav Eirich Award 2011 (Germany), American Ceramic Society Spriggs Phase Equilibria Award 2012 (MS&T'12), URS Pyramid Award 2012 in innovation (USA), TMS Best Professional Energy Paper Award 2017 (TMS2018), and IChemE Global Award 2018 for energy (UK). He serves/served on the Technical Advisory Committee for MOLTEN; International Advisory Board for CIMTEC; TMS Advanced Characterization, Testing, and Simulation Committee; TMS Alloy Phases Committee; organizer for the Advanced Real-Time Imaging Symposium (TMS); and a guest editor for *JOM*.

P. Chris Pistorius is POSCO Professor and Co-Director of the Center for Iron and Steelmaking Research in the Department of Materials Science and Engineering at Carnegie Mellon University. His researches focus on pyrometallurgy and solidification. Previously, he was Associate Professor (1991–1996) and Professor (1997–2008) in the Department of Materials Science and Metallurgical Engineering, University of Pretoria, South Africa, and served as Head of the Department from May 2002 to June 2008. He has a master's degree in metallurgical engineering from the University of Pretoria and a Ph.D. from the University of Cambridge, UK.

Candan Tamerler joined the Department of Mechanical Engineering, University of Kansas, in 2013. She was previously Research Professor in the Materials Science and Engineering Department and Assistant Director of the Genetically Engineered Materials Science and Engineering Center, an NSF MRSEC, at the University of Washington. Prior to that, she was Full Professor at Istanbul Technical University and also served as Chair of the Molecular Biology and Genetics Department at ITU between 2002 and 2010, concurrently holding Visiting Professor position in the Materials Science and Engineering Department at the University of Washington.

Her research interests are in molecular biomimetics, bio-nanotechnology, and bio-enabled materials science. She has published over 90 refereed journal articles, 30 conference proceedings papers, as well as five book chapters and patents. She has organized several symposia and conferences including those sponsored by the American Chemical Society, The Minerals, Metals & Materials Society (TMS), and the Materials Research Society. She has been awarded visiting scientist and professorship positions at the University of Westminster and the University of Nagoya, respectively, and is Principal Member of the Turkish Academy of Sciences.

Hideyuki Yasuda is Professor in the Department of Materials Science and Engineering, Kyoto University. He received his Doctor of Engineering from Kyoto University in 1991. Since then, he had been Assistant Professor, Associate Professor, and Professor in the Graduate School of Engineering, Osaka University (1991–2013). He has been at Kyoto University since 2013. His research fields are solidification and casting of metallic alloys and ceramics. Recently, he has studied time evolution of solidification structure and casting defects by using 2D/3D X-ray imaging in SPring-8. He has also developed modeling and simulation of solidification phenomena on the basis of in-situ observations.

Zuotai Zhang, Ph.D. from Royal Institute of Technology, Sweden, currently holds the position of Professor of the School of Environmental Science and Engineering at Southern University of Science and Technology (SUSTech), and Director of Key Laboratory of Municipal Solid Waste Recycling and Management of Shenzhen City. Before joining SUSTech, he was an associate professor of the Department of Energy and Resources Engineering at Peking University. Dr. Zhang's research interests mainly focus on the efficient and clean utilization of solid waste. He has conducted research on the integrated utilization of waste heat/resource of metallurgical slag, the recycling valuable elements of metallurgical slag, the transformation and utilization of energy solid waste, and the transformation and migration mechanism of harmful elements based on key scientific issues in the process of the efficient cleaning utilization of solid waste. He has published more than 120 articles in leading journals including *Nature Communications*, *Bioresources Technology*, *Journal of Hazardous Materials*, *ACS Sustainable Chemistry & Engineering*, and *Waste Management*, and he has applied for 30 patents. Dr. Zhang also serves on the Board of Review of *Metallurgical and Materials Transactions B* and serves as Vice-Chair of the Solid Waste Subcommittee of the Chinese Ceramic Society of China.

Neslihan Dogan is Assistant Professor of Materials Science and Engineering, and Stelco Research Chair on Sustainable Steel Production at McMaster University. She has bachelor's degree from Yıldız Technical University (Turkey) and Ph.D. from Swinburne University of Technology (Australia). She worked as Postdoctoral Fellow at the University of Wollongong (Australia). She joined the Department of Materials Sciences and Engineering at McMaster University in September 2013.

Her research interest lies in the area of process metallurgy. Her research focuses on the kinetics and thermodynamics of chemical reactions in the steelmaking process using modeling and high-temperature experimental techniques. Her team is currently investigating ways to optimize the current practices during steel refining processes and new techniques to produce the third-generation steel to help steelmakers address some of the issues in environmental and economic sustainability such as energy intensity, material efficiency, and alternative production methods.

Wanlin Wang is Lotus Scholar Chair Professor and Dean of the National Center for International Research of Clean Metallurgy, China. Also, he is Associate Dean of the School of Metallurgy and Environment, Central South University. He received his bachelor's degree in metallurgical engineering from Central South University, master's degree in metallurgical engineering from the University of Utah, and Ph.D. in materials science and engineering from Carnegie Mellon University in 2007. He worked as Senior Research Scientist for Reckitt Benckiser, North America, before joining Central South University in 2009.

His research focuses on high-temperature processing and continuous casting technology. He is the author of more than 130 publications in prestigious international journals, including 45 in *Metallurgical and Materials Transactions B*. He is the recipient of many academic honors, including AIME-Champion H. Mathewson Award from TMS, Marcus A. Grossmann Young Author Award from ASM, Newton Advanced Fellow Award from Royal Society of UK, Excellent Young Scholar Fellowship from National Science Foundation of China, LMD Energy Best Paper Award—Professional from TMS, Kim$^+$—MMI Best Paper Award from Korean Institute of Metals, Weishoukun Scientific Award, and Chinese New Century Excellent Talent Program Award. He is a distinguished member of The Chinese Society of Metals. He has been invited to be an advisory board member and give keynote speeches for numerous international conferences, such as International Congress on the Science and Technology of Steelmaking and International Conference on Molten Slags, Fluxes and Salts.

Noritaka Saito is Associate Professor in the Department of Materials Science and Engineering at Kyushu University, Japan. After earning his Ph.D. from Kyushu University in 2005, he received a Postdoctoral Fellowship from the Japan Society for the Promotion of Science (JSPS), and in 2008, he joined the Department of Materials Science and Engineering at the University of California, Berkeley as Visiting Scholar. In 2009, he joined Kyushu University as Associate Professor with a focus on measurements of physicochemical properties of oxide-based melts at high temperature mostly related to iron- and steelmaking. He has published more than 90 articles in refereed journals and conference proceedings and has been awarded five patents. His current research includes detection and quantification of the solid and gas phases dispersed in oxide melts characterized under alternative current field at high temperatures, and understanding apparent viscous behavior of suspension and forming slag melts. He was the recipient of the TMS Young Leaders International Scholar Award in 2012 and the Sawamura Best Paper Award from the Iron and Steel Institute of Japan in 2014 and 2017.

Bryan Webler received a B.S. in engineering physics from the University of Pittsburgh in 2005 and M.S. (2007) and Ph.D. (2008) in materials science and engineering from Carnegie Mellon. His Ph.D. work at CMU was with the Center for Iron and Steelmaking Research (CISR), studying high-temperature oxidation of low-carbon steels. From 2008 to 2013, he was Senior Engineer in the Materials Technology Department of the Bettis Laboratory, studying corrosion resistance and mechanical behavior of stainless steels and nickel-base alloys. He joined CMU as Assistant Professor in 2013 and became Associate Professor in 2018. He is currently a faculty member in CISR. His research interests are in thermodynamics and kinetics of reactions and transformations in metals. These fundamentals can be applied to a variety of problems in metal processing such as liquid steel refining and microstructure development during additive manufacturing, and metal in-service performance such as high-temperature oxidation and aqueous corrosion.

Part I
Iron & Steelmaking I

A Novel Method of Surface Tension Test for Melt Slags Based on Hot Thermocouple Technique

Zhe Wang, Guanghua Wen, Ping Tang and Zibing Hou

Abstract Currently, the sessile drop method and the maximum bubble method are widely used to measure the surface tension of melt slags. For the above methods, the slow heating rate of furnace decreases the test efficiency. The hot thermocouple technique is widely used in the measurement of the high-temperature performance of molten slags. The slag will form droplets on the thermocouple due to the capillarity, and based on the rapid heating-up and Young's equation, this study used the Single Hot Thermocouple Technique (SHTT) for the test of $CaO–SiO_2–Al_2O_3$ (CSA) and the $CaO–SiO_2–Al_2O_3–MgO$ (CSAM) slag systems. The results show that the interfacial tension between the CSA, CSAM slags and the thermocouple is 2117.76–2131.89 mN/m at 1500 °C. The surface tension of the CSA, CSAM slags can be obtained by Young's equation. Compared with the surface tension measured by the standard test, the SHTT surface tension test error is within 5%.

Keywords SHTT · Molten slag · Contact angle · Interfacial tension
Surface tension

Introduction

In the steel industry, surface tension is one of the important properties of slag, which affects the interface reaction with steel, the inclusions absorption, the foaming of slag, etc. Currently, the recognized methods of surface tension test of slags include the maximum bubble method, the sessile drop method, and the pendant drop method.

In 1917, Jaeger [1] introduced the maximum bubble method into the high-temperature melt to indicate the tensile test. The maximum bubble method need not measure the contact angle, has high accuracy, and is often used for dynamic detection of high-temperature melt surface tension. In 1869, Quincke [2] invented

Z. Wang · G. Wen (✉) · P. Tang · Z. Hou
College of Materials Science and Engineering,
Chongqing University, Chongqing 400044, People's Republic of China
e-mail: wengh@cqu.edu.cn

J. Nakano et al. (eds.), *Advanced Real Time Imaging II*, The Minerals, Metals & Materials Series, https://doi.org/10.1007/978-3-030-06143-2_1

the drop method. The method calculates surface tension based on the shape of the droplet formed by the melt on the horizontal gasket. Due to the small sample size of the sessile drop method and the unrestricted test sample, it is widely used in melt surface tension testing in various fields [3]. In 1882, Bashforth and Adams proposed the pendant drop method [4]. This method is also a simple and applicable method for measuring the surface tension of melts.

Although the above methods for measuring the surface tension of molten slag have been widely recognized, the experiment has the disadvantage of being time-consuming. The maximum bubble method requires a sample bath, and the process of heating the silicon–molybdenum furnace to 1500 °C and completely melting the sample takes about 3 h. The sessile drop method and the pendant drop method melt faster, but the test environment needs to be raised to the required high temperature (e.g., 1500 °C), which takes about 3 h. Therefore, the traditional test method cannot quickly test the surface tension of the slag.

The Single Hot Thermocouple Technique (SHTT) is a kind of high-temperature measurement method which uses intermediate frequency chopper technology to realize both heating and temperature measurement. The SHTT method has a small sample volume, a fast heating rate (0–30 °C/s) [5, 6], a high temperature rise temperature (1800 °C) [7], a fast cooling rate (about 150 °C/s) [8], and other advantages. So the SHTT method is widely used in glass, ceramics, and other fields of melting, continuous cooling crystallization, and constant temperature crystallization [8–10]. Fred Ordway [9] pointed out that the slag will form droplets on the thermocouple of SHTT due to capillary action. Wang, Zhou et al. further proposed that the contact angle of the slags' droplets with thermocouple can be obtained by SHTT [11] and combining the Double Hot Thermocouple Technique (DHTT) and Young's equation can get the interfacial tension of slag-metal [12]. But they lack the comparison of the reliability of the above methods and the study of the practicality of specific metallurgical slag systems.

According to Young's equation [13], when the contact angle and surface free energy of metal are known, if the interfacial tension between slag and gold is constant, the surface tension of molten slag can be calculated. Jung and Min [14], in the study of the influence of Al_2O_3 and MgO on the interfacial tension between CSAM slag and solid steel, pointed out that the composition change has little effect on the interfacial tension of slag-metal. From this, it can be inferred that if the slag-metal interfacial tension is constant, the surface tension of the CSA and CSAM slag can be obtained by the SHTT method.

Based on the characteristics of rapid heating by SHTT method and the capillary action between slag and thermocouple, the feasibility of the above conjecture was verified by experiments. Furthermore, a novel method for measuring the surface tension of CSA and CSAM slags based on SHTT was proposed.

Table 1 Chemical compositions of all samples in (wt%)

No.	CaO	SiO_2	Al_2O_3	MgO	References
1	35.00	60.00	5.00	0.00	[15]
2	40.00	40.00	20.00	0.00	[15]
3	37.50	37.50	25.00	0.00	[15]
4	37.60	38.00	16.50	7.90	[16]
5	42.80	37.10	15.20	4.90	[16]

Materials and Methods

Sample Preparation

The iron and steel metallurgical slag is based on $CaO–SiO_2–Al_2O_3–MgO$. According to different production process requirements, the flux is added to meet the requirements of metallurgy process. In this study, $CaO–SiO_2–Al_2O_3–MgO$ (CSAM) and $CaO–SiO_2–Al_2O_3$ (CSA) slag samples were selected for experiments. The components of sample 1–5 and the references are shown in Table 1.

Analytical reagents are used for the preparation of samples. In the process of preparation, the samples were first put into the graphite crucible and homogenized for 30 min in a silicon–molybdenum resistance furnace at 1400 °C; then the homogenized slag was poured into a water-cooled copper plate to quench the slag. The slag sample was subjected to XRD analysis to ensure that the slag sample was uniform and amorphous; finally, the quenched slag sample was placed in an agate crucible and passed through a 200-mesh sieve to obtain a surface tension test slag sample.

SHTT Method

The schematic diagram of the SHTT device is shown in Fig. 1. Kashiwaya has explained the principles of SHTT in detail [5]. Based on the SHTT method, a kind of furnace with controllable atmosphere and auxiliary heat source is designed by our research group. The sealed design of the furnace body can ensure the stability of the test environment, and the atmospheres can be set according to the experimental requirements. Auxiliary heating uses a high-temperature halogen lamp as a heat source to provide thermal compensation for the thermocouple quickly. Therefore, the SHTT equipment can be used to test the interface properties of molten slag under controlled atmosphere and stable ambient temperature.

Experimental Procedure

The experiment analyzes the properties of the slag interface which bases on Young's equation, as shown in Eq. 1. In Eq. 1, γ_s is the surface free energy of platinum, which is 2370 mN/m [17], γ_{sl} is the slag-metal interfacial tension, γ_l is the surface tension of slag, and θ is the contact angle of slag and platinum–rhodium alloy thermocouple.

$$\gamma_s = \gamma_{sl} + \gamma_l \times \cos\theta \tag{1}$$

The SHTT method for the surface tension test of CSA and CSAM slag: (i) chooses air atmosphere as the test atmosphere for the surface tension test of molten slag; (ii) increases the temperature of the thermocouple to 1500 °C and keeps the temperature of the thermocouple; (iii) the sample is in contact with the high-temperature thermocouple and melted; (iv) the droplet size is adjusted to the appropriate size (about 1 mg) and makes the central axis of the droplet parallel to the observation surface; (vii) bringing the platinum surface free energy, contact angle, and slag surface tension literature values into Young's equation to calculate the slag-metal interfacial tension; (viii) if the experimental slag-metal interfacial tension varies little with composition, the average interfacial tension is used as a constant and substituted into Young's equation to calculate the surface tension of the slag; (ix) the calculated surface tension of the slag is compared with the literature value to verify the accuracy of the surface tension of the CSA, CSAM slag based on SHTT.

In step (vi), to ensure the accuracy of the contact angle data, the contact angle measurement is calculated according to the Laplace equation, as shown in Eq. 2 [18]. The droplet profile is first fitted as a circle, then the height "*h*" of the droplet on the upper surface of the thermocouple is measured, and the contact radius "*a*"

Fig. 1 Schematic of SHTT

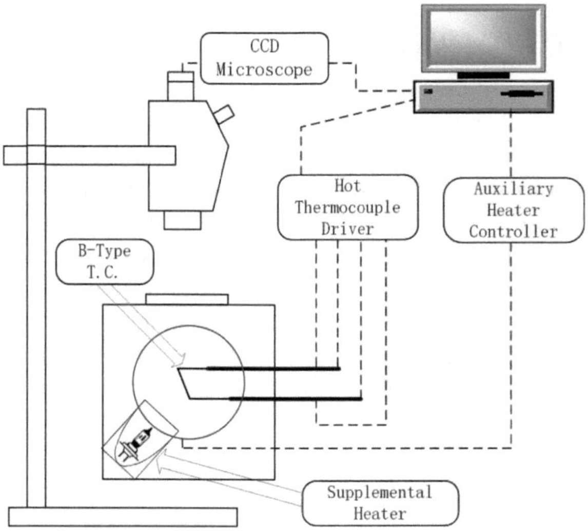

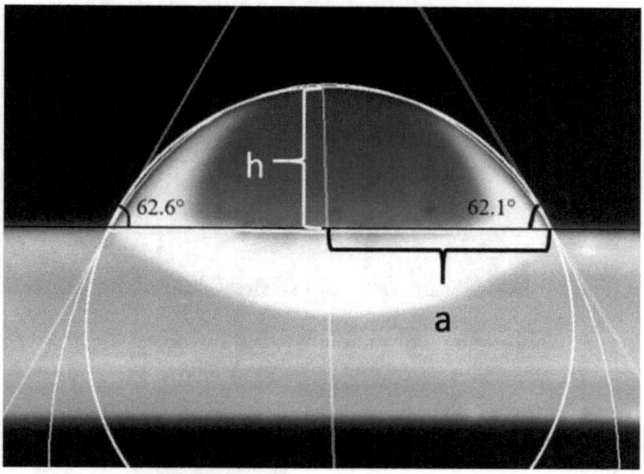

Fig. 2 Schematic of contact angle test

between the droplet and the upper surface of the thermocouple is measured. Finally, the contact angle is obtained by using the Laplace equation, as shown in Fig. 2. Based on the above principle, the contact angle plug-in of ImageJ software was used in this study to measure the contact angle.

$$\tan\left(\frac{\theta}{2}\right) = \frac{h}{a} \tag{2}$$

Results and Discussion

Contact Angle Test

The SHTT surface tension test method is based on the contact angle between the slag and the Pt–Rh thermocouple, and Young's equation is used to calculate the surface tension of the slag. Therefore, the contact angle is the key point of the SHTT slag interface property test method, which directly affects the accuracy of the test results. In Young's equation, the contact angle is defined as the contact of the droplet with the horizontal gasket. To explore the rationality of the contact angle obtained by the SHTT test method applied to Young's equation, the sample 2 was selected for the sessile drop method.

In the sessile drop method, the atmosphere was air, and the substrate was platinum. In the experiment, the molten of sample 2 dropped on the platinum substrate in a furnace which was 1500 °C and then kept at the same temperature for 5 min. When the shape of the droplets did not change, the droplet was photographed using a camera, and the contact angle was measured by ImageJ software; the results were as shown in Fig. 3. From Fig. 3, the result of the sessile drop method experiment is 60.00°, which is similar to the SHTT method test result of 60.12°,

Fig. 3 Contact angle of sample 2 tested by the sessile drop method

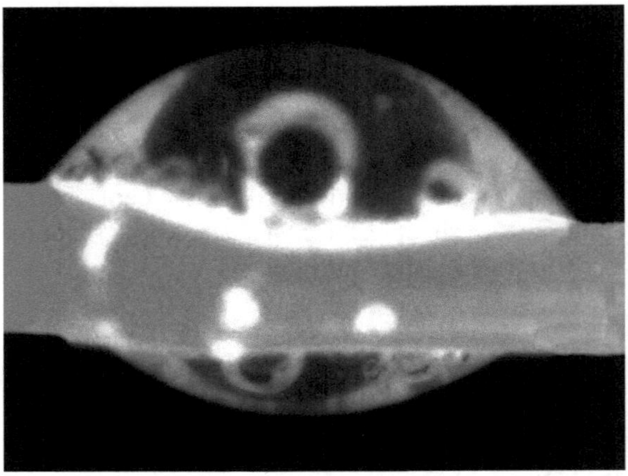

Fig. 4 Incorrent SHTT contact angle test result diagram

and the relative error is 0.20%. Therefore, the contact angle obtained by SHTT can satisfy the calculation of Young's equation.

To ensure the accuracy of the contact angle test, the test procedure must comply with the following four requirements. (i) Ensure that the Pt–Rh thermocouple test area is straight because the curved thermocouple directly affects the contact angle numerical accuracy. (ii) The size of the droplet should be suitable (about 1 mg). If the droplet is too large, the droplet will flow under the Pt–Rh thermocouple due to gravity, as shown in Fig. 4. (iii) Excluding the bubbles in the droplets. The bubbles in the droplets will affect the contact angle test results, as shown in Fig. 4. (iv) Ensure the axis of the droplet is parallel to the observation surface, so that the SHTT slag-metal contact angle test angle is consistent with the sessile drop method.

To verify the reproducibility of the SHTT contact angle test, the study selected sample 1 and sample 5 for reproducibility experiments. According to the above test specifications, each group of slag was tested three times and averaged, as shown in Table 2. It can be seen from Table 2 that the maximum fluctuation range of the contact angle of the reproducibility experiment is $\pm 2.00°$, so the SHTT experiment can meet the contact angle reproducibility requirement.

Table 2 Contact angle reproducibility test data

No.	Contact angle (°)	Average contact angle (°)	Angle fluctuation range (°)
1	56.00	54.00	±2.00
	54.00		
	52.00		
5	63.60	62.78	±1.15
	61.30		
	63.43		

Table 3 Slag-metal interfacial properties

No.	Contact angle (°)	γ_l (mN/m)	γ_{sl} (mN/m)
1	54.00	418.70	2123.89
2	60.12	477.96	2131.89
3	64.50	585.92	2117.76
4	61.03	496.50	2129.48
5	62.78	543.30	2121.45

Interfacial Tension

In order to get the surface tension of slag by SHTT, the interfacial tension between slag and Pt–Rh thermocouple should be kept stable. It can be known from Young's equation that the interfacial tension between slag and metal can be calculated when the surface tension and contact angle of slag are known. The contact angle measured by SHTT and the slag surface tension value in the literature were substituted into Young's equation, and the slag-metal interface tension was calculated, as shown in Table 3. From Table 3, the slag-metal interfacial tension values of sample 1–5 remain stable, the range of change is 2117.76–2131.89 mN/m, and then, the interfacial tension between CSA, CSAM slag and thermocouple remains stable. The experimental results are consistent with the results of Jung and Min [14]; that is, the change of CSAM slag composition has little effect on the interfacial tension of slag-metal.

Surface Tension

It is conjectured that if the interface tension between slag and metal is constant, the contact angle and the free energy of platinum surface can be substituted into Young's equation to calculate the surface tension of molten slag. In the section "Interfacial Tension," the calculation results of interfacial tension show that the interfacial tension of samples 1–5 is less affected by the change of composition. Therefore, the average interfacial tension of samples 1–5 (2124.89 mN/m) is regarded as a constant, and the calculated value of the surface tension of the slag is obtained by

Table 4 Comparison of literature values and calculated values of slag surface tension

No.	γ_l literature value (mN/m)	γ_l calculated value (mN/m)	Relative error (%)
1	418.70	416.99	0.41
2	477.96	492.01	2.94
3	585.92	569.35	2.83
4	496.50	505.98	1.91
5	543.30	535.78	1.38

Young's equation. The surface tension calculation values and literature values were further compared to explore the feasibility of calculating the surface tension of CSAM slag based on the SHTT method.

After calculation, the calculated surface tension of samples 1–5 at 1500 °C was obtained, as shown in Table 4. From Table 4, the relative error between the calculated value of the surface tension of samples 1–5 and the literature value is 0.41–2.94%. Therefore, based on the SHTT, the surface tension of CSA and CSAM slag can be calculated relatively accurately.

Conclusion

In order to verify the feasibility of testing the surface tension of molten slag based on SHTT, $CaO–SiO_2–Al_2O_3–MgO$ and $CaO–SiO_2–Al_2O_3$ metallurgical slag were selected in this study. The contact angle between slag and B-type thermocouple, the interfacial tension between slag and metal, and the surface tension of slag were studied. The main conclusions are as follows:

(1) The contact angle between slag and B-type thermocouple was measured by SHTT method. The results show that slag can be absorbed to the thermocouple and form a stable droplet on the thermocouple because of the capillary effect. Under the same conditions, the relative error of the contact angle test results with the sessile drop method was 0.20%, and the fluctuation range of the SHTT contact angle repeated test was $\pm 2.00°$. Then the contact angle obtained by the SHTT can satisfy Young's equation.

(2) The interfacial tension test between the slag and the B-type thermocouple shows that the interfacial tension value changes little with the slag composition. The average value of the slag-metal interfacial tension was taken as a constant into Young's equation. The relative error between the calculated value of the surface tension of the molten slag and the reference value is 2.94%. The SHTT-based slag surface tension test method can accurately determine the surface tension of $CaO–SiO_2–Al_2O_3$, $CaO–SiO_2–Al_2O_3–MgO$ slag.

Acknowledgements The authors greatly appreciate the funding from the National Natural Science Foundation of China (Grant No. 51574050).

References

1. Boni RE, Derge G (1956) Surface tensions of silicates. JOM 8:53–59
2. Quincke G (1869) Ann Phys 138:141
3. Fujii H et al (2006) Surface tension of molten silicon measured by microgravity oscillating drop method and improved sessile drop method. Acta Mater 54:1221–1225
4. Bashforth F, Adams J (1883) An attempt to test the theories of capillary action. University Press, Cambridge, England
5. Kashiwaya Y et al (1998) Development of double and single hot thermocouple technique for in situ observation and measurement of mold slag crystallization. ISIJ Int 38:348–356
6. Zhou L et al (2012) A kinetic study of the effect of basicity on the mold fluxes crystallization. Metall Mater Trans B 43:354–362
7. Welch JH (1954) A simple microscope attachment for observing high-temperature phenomena. J Sci Instrum 31:458
8. Kashiwaya Y, Cicutti CE, Cramb AW (1998) An investigation of the crystallization of a continuous casting mold slag using the single hot thermocouple technique. ISIJ Int 38:357–365
9. Ordway F (1952) Techniques for growing and mounting small single crystals of refractory compounds. National Bureau of Standards
10. Wang W et al (2018) High-temperature properties of mold flux observed and measured in situ by single/double hot-thermocouple technique. JOM 70:1248–1255
11. Zhou L et al (2018) Effect of Li_2O on the behavior of melting, crystallization, and structure for $CaO–Al_2O_3$-based mold fluxes. Metall Mater Trans B 49:2232–2240
12. Zhou L et al (2013) Computational modeling of temperature, flow, and crystallization of mold slag during double hot thermocouple technique experiments. Metall Mater Trans B 44:1264–1279
13. Cramb AW, Jimbo I (1989) Calculation of the interfacial properties of liquid steel–slag systems. Steel Res 60:157–165
14. Jung EJ, Min DJ (2012) Effect of Al_2O_3 and MgO on interfacial tension between calcium silicate-based melts and a solid steel substrate. Steel Res Int 83:705–711
15. Li Y et al (2011) Determination and calculation for the surface tension of $CaO–Al_2O_3–SiO_2$ melts. Male, Maldives
16. Allibert M et al (1995) Slag Atlas, 2nd edn. Verlag Stahleisen GmbH, Düsseldorf, p 446
17. Tran R et al (2016) Surface energies of elemental crystals. Sci Data 3:160080
18. Butt HJ, Graf K, Kappl M (2006) Physics and chemistry of interfaces. Wiley

In Situ Observation on the Interactions of Nonmetallic Inclusions on the Surface of Liquid Steel

Youngjo Kang, Piotr R. Scheller, Du Sichen and Kazuki Morita

Abstract The behaviors of several types of inclusions on the surface of liquid steel were examined using a confocal laser scanning microscope, CLSM. While alumina inclusions are likely to attract each other, agglomerate, and grow fast by the agglomeration, the other types of inclusions such as spinel and calcium aluminate rarely interact with each other. The analysis on the observation result of the CLSM indicates that the attraction force and the agglomeration play a significant role in the growth of alumina inclusions. Moreover, the behaviors of liquid calcium aluminate inclusions, which were intentionally injected, could be carefully observed. Their agglomeration took place only when they occasionally collide under the existence of external force, in spite of relatively low collision probability.

Keywords Nonmetallic inclusion · Alumina · Spinel · Agglomeration
Confocal scanning laser microscope

Y. Kang (✉)
Department of Materials Science and Engineering, Dong-a University,
Busan 49315, Korea
e-mail: youngjok@dau.ac.kr

P. R. Scheller
Freiberg University of Mining and Technology, 09596 Freiberg, Germany

D. Sichen
Department of Materials Science and Engineering, Royal Institute of Technology,
100 44, Stockholm, Sweden

K. Morita
Department of Materials Engineering, The University of Tokyo, Tokyo 113 8656, Japan

J. Nakano et al. (eds.), *Advanced Real Time Imaging II*, The Minerals,
Metals & Materials Series, https://doi.org/10.1007/978-3-030-06143-2_2

Introduction

Although nonmetallic inclusions in steel often impose harmful effect on the product quality, they usually float in molten steel and are possibly removed into top slag because of density difference. Obviously, the efficiency of floatation and removal of inclusions is dependent on the size of inclusions. It has been also known that the agglomeration by random collisions in molten steel is one of the most important mechanisms of physical growth of inclusions [1]. Therefore, it will be greatly necessary to understand about the mechanism of inclusion growth by agglomeration for proper inclusion control.

Since the observation technique using laser scanning confocal microscope was innovatively brought for a direct examination on the behavior of inclusions, [2] a few more notable researches of in situ observation about the behaviors of various type of inclusions have been subsequently made [3–6]. In spite of meaningful and intensive discussions in those works, however, more explanation is required in the unclear phenomena, such as different interactive tendencies of each type inclusion. In this research, the behaviors of various kinds of inclusions on the molten steel surface were examined using a laser scanning confocal microscope. In situ observation on the agglomeration behaviors of each inclusion was discussed.

Experimental

In situ observation was carried out using a laser scanning confocal microscope (1LM21H/SVE17SP, Lasertec), which consists of He–Ne laser source, a detector, video recording unit, and a high temperature image furnace. Figure 1 schematically shows the gold-coated image furnace, where sample can be heated by radiation from a halogen lamp in the bottom. The sample can be observed through an observation window equipped on the top of the furnace.

A steel sample, which was prepared in advance to fit in a small alumina crucible (ϕ 9 mm, h 4 mm), was placed on a platinum sample holder. During heating, temperature was controlled by PID temperature control program and R-type thermocouple placed under the platinum holder. Various atmospheric conditions in the furnace were chosen to allow the formation of various types of inclusions, such as Al_2O_3, $MgO \cdot Al_2O_3$, and $CaO–Al_2O_3$. For the better observation of calcium aluminate liquid inclusions, synthetic inclusions with a typical composition ($12CaO \cdot 7Al_2O_3$) were prepared by premelting and pulverizing into fine powder under 38 μm. The inclusion powders were inserted between two steel disks after thoroughly mixing with far larger amount of iron powders (~150 μm). After the observation, the inclusions on the surface of sample were identified using SEM (Leo 430, Leo Electron Microscopy) and EPMA (JXA-8800, JEOL) (Table 1).

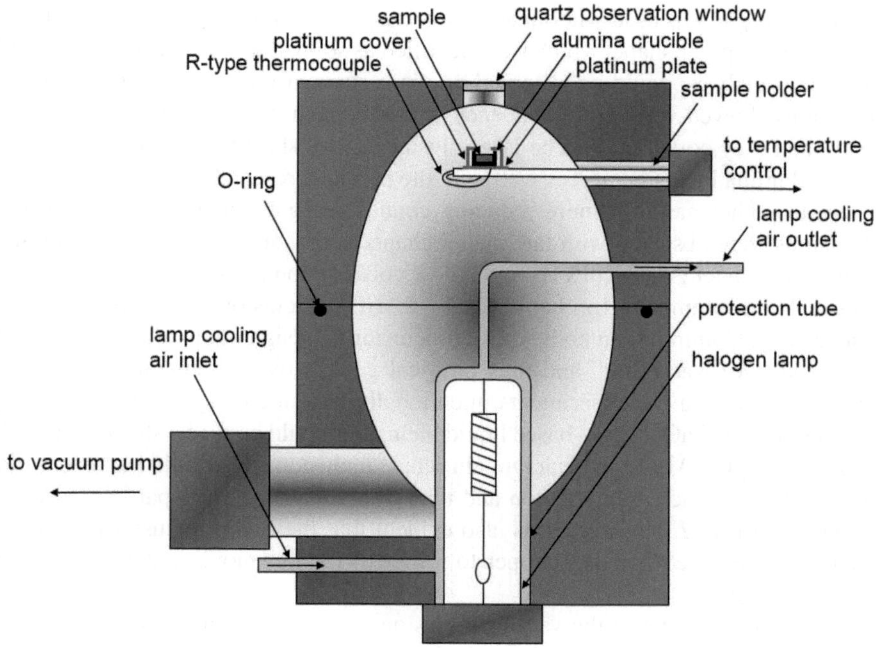

Fig. 1 Schematic diagram of an image furnace of laser scanning—confocal laser scanning microscope

Table 1 Typical chemical compositions of steel in the present study

Element	C	Si	Mn	Cr	Mo	V
wt%	0.39	1.0	0.4	5.3	1.3	0.9

Results and Discussion

As expected, different types of inclusions could be observed under the different atmospheres. Observed inclusions under each atmospheric condition are summarized in Table 2.

Table 2 Types of inclusions under various atmospheric conditions

Atmospheric conditions	Observed inclusions
Ar ($O_2 \leq 2$ ppm)	Mostly Al_2O_3 (endogenous)
Ar + 1 vol% H_2	Mostly $MgO \cdot Al_2O_3$ (endogenous)
Ar + 2 vol% H_2	$CaO \cdot 2Al_2O_3$, $MgO \cdot Al_2O_3$ (endogenous)
Ar + 3 vol% H_2	Al_2O_3 (endogenous), liquid CaO–Al_2O_3 (intentionally introduced)

When Ar gas was used without any reducing gas addition, large number of inclusions with irregular shape, which was identified as solid alumina inclusions by SEM or EPMA analysis, could be observed on the surface of molten steel. The alumina inclusions showed very strong tendency to attract each other and agglomerated so easily that they could grow to be large cluster in very short time. This observation on the alumina inclusion agrees well with the previous researches [2].

On the other hand, in more reducing conditions, polygonal or globular small inclusions were observed with far smaller chances compared to the case of alumina inclusions. Under Ar + 1 vol%H_2 and Ar + 2 vol%H_2 conditions, $MgO \cdot Al_2O_3$ and solid calcium aluminate (probably $CaO \cdot 2Al_2O_3$) inclusions could be detected, respectively. Commonly in both types of inclusions, no sign of attractive interaction (nor repulsive) was found, and thus physical growth by agglomeration seems to be insignificant. In the experiments intentionally introducing synthetic inclusions, large number of $CaO–Al_2O_3$-based liquid inclusions could be successfully observed. Similar to $MgO \cdot Al_2O_3$ and calcium aluminate inclusions, the liquid inclusions did not show any attractive interaction and tend to maintain the inter-particle distance as shown in Fig. 2. However, it is also evident that the liquid inclusions could be coalesced with ease when they happen to collide by random movement on the surface of molten steel.

Different behaviors of the various inclusions could be confirmed by the change in the inter-particle distance as shown in Fig. 3. Contrary to other inclusions, which have no interactions on the molten steel surface, strong attraction between alumina inclusions was found to be effective even though the inclusion pair was initially departed from each other more than 100 μm. Based on the acceleration and the sizes of the inclusions, moreover, the force acting on the inclusion pairs could be in the range of 10^{-18} to 10^{-14} N.

To properly reveal the origin of the attractive force of various inclusions, the universal gravitation and the capillary forces were evaluated for several pairs of inclusions observed under the microscope. As indicated in Fig. 4, the gravitation force

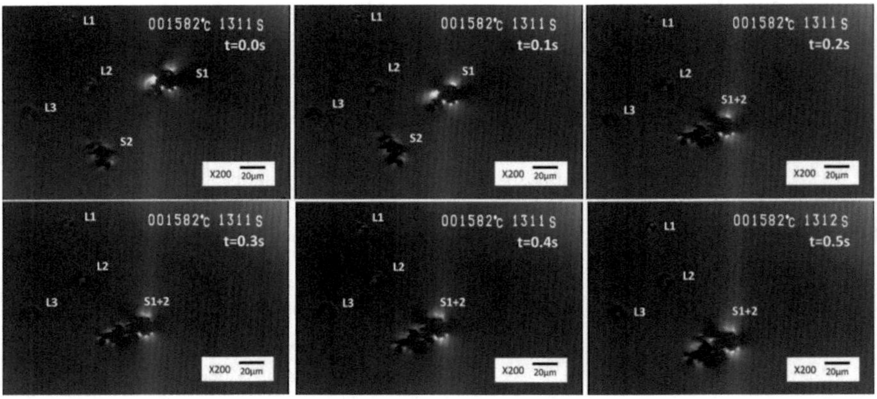

Fig. 2 Images for the comparison between the interactions among liquid inclusions and solid inclusions

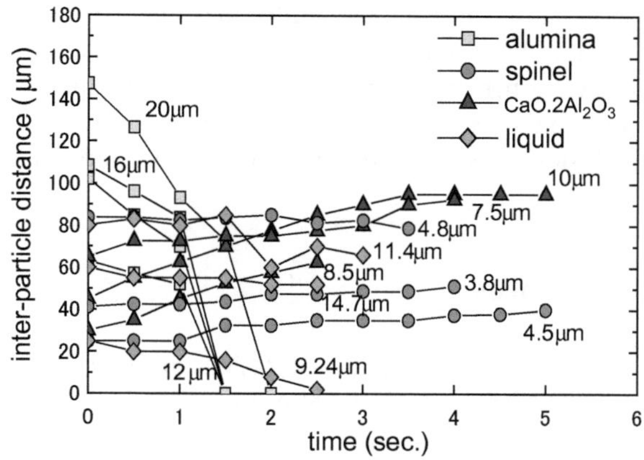

Fig. 3 Inter-particle distance of various types of inclusions as function of time

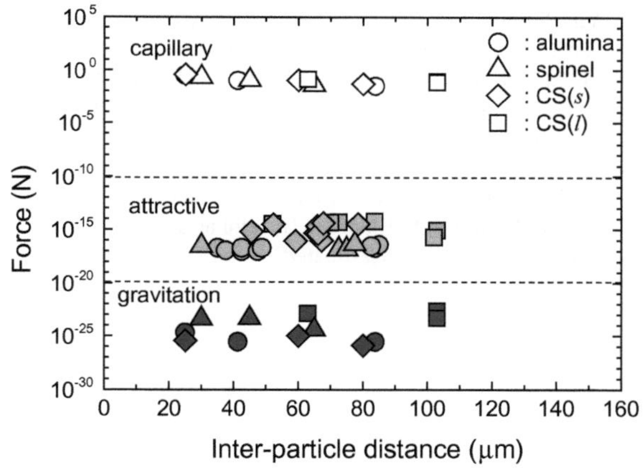

Fig. 4 Forces acting on various inclusions as function of inter-particle distance

turned out to be very small compared to the attractive force between the inclusions, whereas the capillary force at the equilibrium position of molten steel surface between a pair of inclusions parallelly immersed plate is supposed to be larger than 10^{-2} N. For more practical estimation of the capillary force, very careful analysis on the depression depth of the molten steel surface may be required. Also, relatively small difference in the capillary force among each type of inclusions may imply that the capillary force is not mainly responsible for different attractive phenomena, contrary to the previous researches [2, 3]. Since the exceptional attractive behavior of the alumina inclusions is not clear yet, however, more elaborate study should be carried out regarding the interaction of the alumina inclusions.

Summary

For better insight into the physical growth of inclusions in molten steel, the behaviors of various types of inclusions on steel surface were observed by direct in situ observation using an LSCM. Various kinds of inclusions could be observed under different atmospheric conditions. While alumina inclusions show extraordinarily attractive behavior, other inclusions such as spinel and calcium aluminate did not show any interaction, keeping their inter-particle distance. Each tendency of various inclusions was discussed with the attractive forces and possible sources of the force.

References

1. Zhang L, Thomas BG, Wang X, Cai K (2002) Evaluation and control of steel cleanliness—Review. Paper presented at the 85th Steelmaking Conference, Nashville, Tennessee, 10–13 March, 2002
2. Yin H, Shibata H, Emi T, Suzuki M (1997) "In-situ" observation of collision, agglomeration and cluster formation of alumina inclusion particles on steel melts. ISIJ Int 37:936–945
3. Yin H, Shibata H, Emi T, Suzuki M (1997) Characteristics of agglomeration of various inclusion particles on molten steel surface. ISIJ Int 37:946–955
4. Shibata H, Yin H, Yoshinaga S, Emi T, Suzuki M (1998) In-situ observation of engulfment and pushing of nonmetallic inclusions in steel melt by advancing melt/solid interface. ISIJ Int 38:149–156
5. Misra P, Chevrier V, Sridhar S, Cramb AW (2000) In situ observations of inclusions at the (Mn, Si)-killed steel/CaO-Al_2O_3 interface. Metall Mater Trans B 31B:1135–1139
6. Colletti B, Vantilt S, Blanpain B, Sridhar S (2003) Observation of calcium aluminate inclusions at interfaces between Ca-treated, Al-killed steels and slags. Metall Mater Trans B 34B:533–538

Part II
Energy, Fuels, and Environment

In Situ Structural Variations of Individual Particles of an Al_2O_3-Supported Cu/Fe Spinel Oxygen Carrier During High-Temperature Oxidation and Reduction

W. H. Harrison Nealley, Anna Nakano, Jinichiro Nakano and James P. Bennett

Abstract Physical and chemical degradation of the oxygen-carrier materials during high-temperature redox exposures may affect the overall efficiency of the chemical looping process. Therefore, studying real-time physical and chemical changes in these materials when exposed to repeated redox cycles is essential for further development of chemical looping technology. In this work, the National Energy Technology Laboratory's Al_2O_3-supported Cu/Fe spinel oxygen carrier, in the form of a $CuO \cdot Fe_2O_3$ solid solution, was examined in situ during 3-h exposures to either oxidizing or reducing environments at 800 °C using a controlled atmosphere heating chamber in conjunction with a confocal scanning laser microscope. A compilation of the physical changes of individual particles using a controlled atmosphere confocal microscope and the microstructural/chemical changes documented using a scanning electron microscope will be discussed.

Keywords Chemical looping · Fossil energy · Oxygen carrier

Introduction and Background

The emission of CO_2 due to carbon-based industrial processes is one of the main contributors to the amount of greenhouse gases in the atmosphere [1]. Measures have been taken to reduce the emissions from sites that produce CO_2 such as power plants,

---

W. H. H. Nealley (✉) · A. Nakano · J. Nakano · J. P. Bennett
U.S. Department of Energy National Energy Technology Laboratory,
1450 Queen Ave, Albany, OR 97321, USA
e-mail: william.nealley@netl.doe.gov

W. H. H. Nealley
Oak Ridge Institute for Science and Education, 100 ORAU Way, Oak Ridge,
TN 37380, USA

A. Nakano · J. Nakano
AECOM, P.O. Box 618, South Park, PA 15129, USA

J. Nakano et al. (eds.), *Advanced Real Time Imaging II*, The Minerals, Metals & Materials Series, https://doi.org/10.1007/978-3-030-06143-2_3

smelters, and gasifiers. Part of the core mechanism of CO_2 emissions lies with the oxidation of carbon in fossil fuels.

If air is used to provide the necessary O_2 to support the combustion process, the presence of N_2, which plays a minimal role in the combustion process relative to O_2, reduces the overall energy efficiency because it must be heated to the process temperature. This also makes subsequent carbon capture of the exhaust more difficult [2].

The carbon capturing would be more efficient if the oxidizer used for combustion produces pure CO_2 exhaust without any other gas species present. Chemical looping combustion can produce pure CO_2 by utilizing oxygen-carrier materials. These materials act by exchanging oxygen during oxidation/reduction (redox) cycles. This technique would be more feasible if using solid oxygen carriers versus those in a liquid or gaseous state pose a number of financial and technical problems [3].

A viable material for this purpose must be resistant to physical and chemical attritions over long periods of time involving high temperatures and a varying gas environment. Multiple metal oxides, both single-metallic and bimetallic, have been tested for their use in this application [4, 5]. Cu and Fe oxides were found to be two of the best candidates due to their ability to oxidize and reduce repeatedly [6]. However, the metal oxide carriers are usually supported on a ceramic, such as Al_2O_3, with which the oxygen carriers of Cu and Fe oxides have been shown to react at chemical looping combustion temperatures [7, 8]. Further improvements in Cu- and Fe-based oxygen carriers would support the material development.

At the National Energy Technology Laboratory (NETL), a Cu–Fe spinel-based material supported on Al_2O_3 is being developed as an oxygen-carrier (OC) candidate for chemical looping applications [9]. The present work continues a study with this material. A previous study discussed part of the material characteristics in high-temperature environments while undergoing repeated redox cycles [3].

Materials and Methods

Material

The OC studied in this work was developed by NETL and manufactured by Nexceris using solid-state mixing. The sample composition was analyzed using X-ray fluorescence spectroscopy (XRF and Rigaku ZSX Primus II) and determined to be 37.49 wt%CuO, 31.32 wt%Fe_2O_3, and 31.19 wt%Al_2O_3. The X-ray diffraction analysis (XRD and Rigaku Ultima IV XRD spectrometer) of as-received OC sample confirmed the presence of two crystalline phases: spinel and alumina. Details of this material can be found in Siriwardane et al. [10].

Confocal Microscopy

For in situ analysis, the samples were placed inside a halogen-lamp furnace (Yonekura) with a quartz observation window, which was attached to a confocal scanning laser microscope (CSLM and Olympus OLS 3100) (Fig. 1). The furnace was also equipped with a gas delivery system.

In situ observations were conducted by first heating a sample to 800 °C at 800 °C/min in synthetic air (79 vol%N_2–21 vol%O_2). Once heating was completed, the air (oxidizing gas) flow was either maintained or switched to a reducing gas mixture of 10 vol%CO–90 vol%Ar (10CO–90Ar hereafter). This was followed by a 5-min hold to stabilize temperature. After the hold, the material's surface was scanned with a 408 nm laser and with white light once every 5 min for the first 30 min of each gas exposure and every 30 min thereafter. 3D data obtained from the scans were used to analyze individual particles' volume change during gas exposures. The gas flow rate was 50 mL/min throughout the test. No particle stirring was performed; that is, sample evaluations were conducted in a static environment other than a gas flowing through the chamber.

The samples were subjected to up to four 180-min gas exposures with cooling to ambient temperature between each exposure (Fig. 2). Due to the extended length of the test, in some cases, samples were cooled to room temperature and stored overnight in the chamber in the synthetic air with no flow before the next gas exposure. Gas exposures were designed to simulate extended redox cycles and followed the progression: oxidation (Ox), oxidation–reduction (Ox–Red), oxidation–reduction–oxidation (Ox–Red–Ox), and oxidation–reduction–oxidation–reduction (Ox–Red–Ox–Red). Following the final gas exposure of each test, the sample was quenched with a

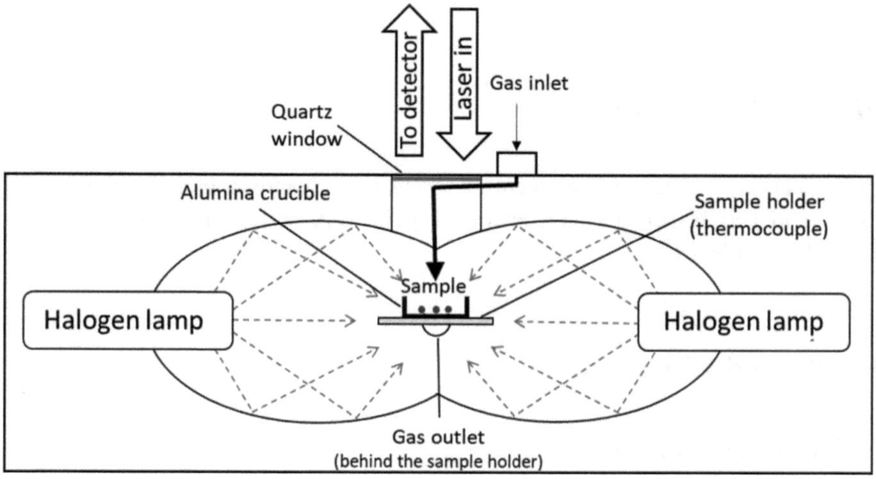

Fig. 1 A schematic of the CSLM heating chamber; reproduced with permission from Nealley et al. [3]

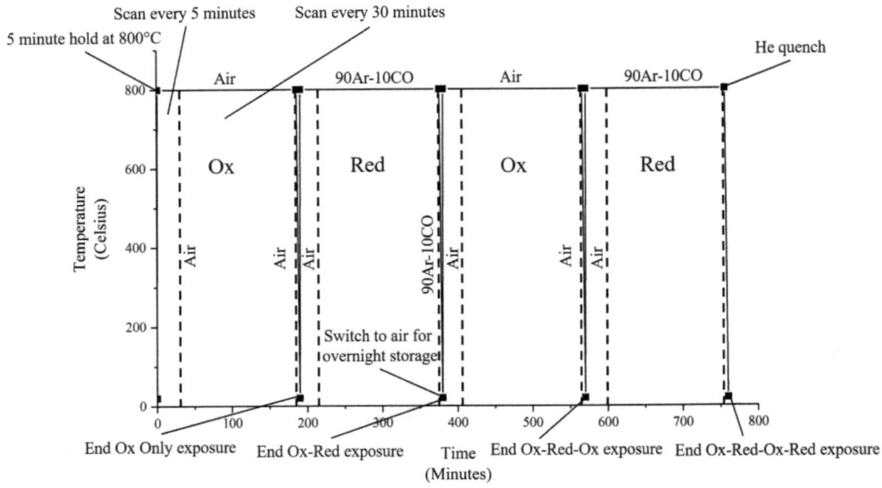

Fig. 2 Example of the experimental temperature profile for the extended gas exposure tests. "Ox"-oxidizing gas; "Red"-reducing gas

60 mL/min flow of He. The samples were then removed from the CSLM chamber and stored in plastic bags inside a desiccator at ambient temperature for post-exposure analysis.

A special procedure for the volume change analysis was developed using Olympus' microanalytical software (LEXT). The uncertainty in the volume measurement was determined to be approximately 1.99%. Since the crucible bottom was not perfectly flat, the volume was measured twice to account for the minor variations. This uncertainty is specific to each data point and reported as error bars (Fig. 3).

Electron Microscopy

Cross-sections of the samples were analyzed by a scanning electron microscope (SEM and FEI Inspect F) with energy-dispersive X-ray spectroscopy (EDX and Oxford INCA Wave).

Results and Discussion

In Situ Observations

Data from laser and white-light scans were used to generate three-dimensional representations of the particles and analyze volume change. For all the tests, during

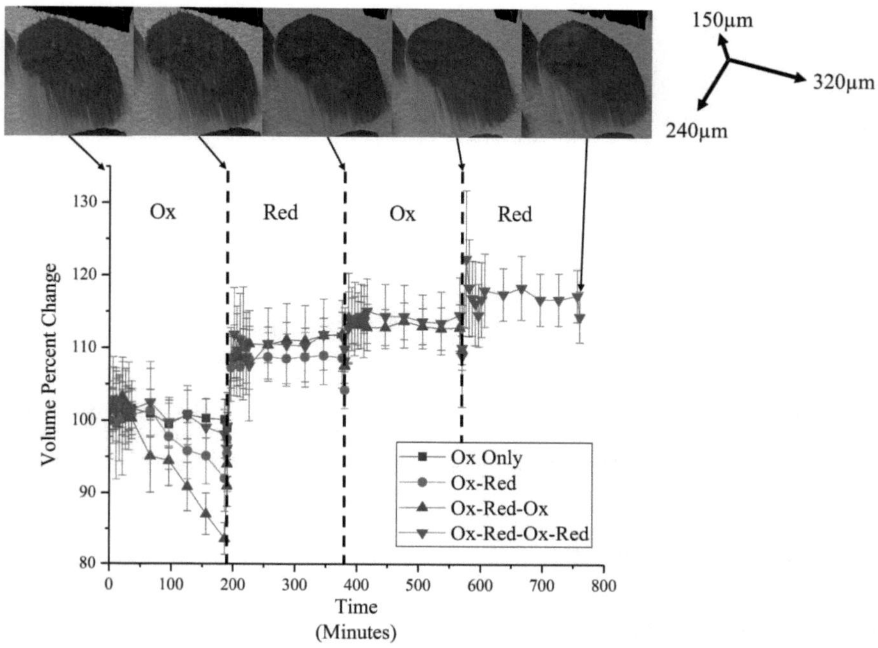

Fig. 3 Volume change measured during exposures relative to the respective particles' initial volumes (100%); the dashed lines indicate the end of one gas exposure and the beginning of another; in situ CSLM images from the Ox–Red–Ox–Red exposure are shown with their corresponding times at 800 °C

the initial oxidation (Ox), the sample tended to decrease in overall volume (Fig. 3). Assuming the original spinel structure was stable under the test conditions during oxidation, volume change was likely caused by sintering at 800 °C. It has been reported Cu/Fe spinel exhibits strong sintering-resistance under similar conditions of air oxidation, 50 vol%H_2 reduction, and 850 °C exposure temperature [11]. Note that the material used in those experiments was prepared via sol–gel combustion synthesis, whereas the present samples were prepared by mechanical mixing [10]. It is possible that the structural differences created by those manufacturing methods would influence the sintering behavior. Furthermore, the material in Wang et al. had already been sintered at 950 °C during the sample preparation procedure [11]. It is not certain whether the sample used for this study was appreciably sintered during the manufacturing process. Note, no sample heating prior to the experimental exposures was performed in this work.

During the following reducing gas exposure (Red), a rapid and substantial jump in volume was noted, indicating that the sample was rapidly altered immediately after exposure to the 10CO–90Ar gas. For the next Ox and Red gas exposures, sample exhibited the step-like volume change with a sharp increase at the beginning of each exposure.

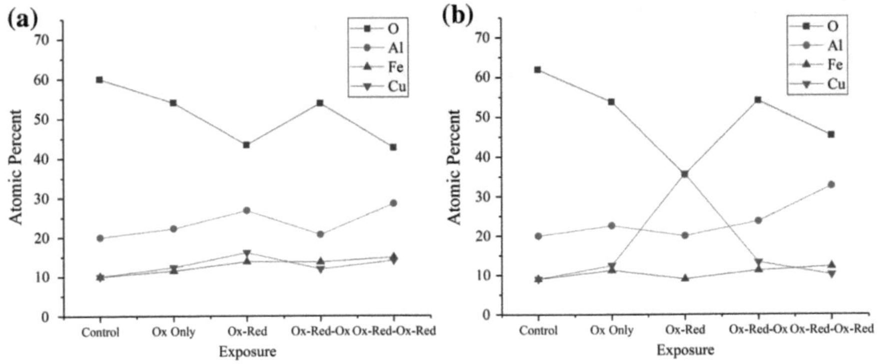

Fig. 4 Average elemental compositions of inner spinel grains plotted over exposure conditions; averages were produced from the five-EDX points taken near the center of the spinel grain; randomly selected spinel grains were measured near the center (**a**) and at the edge (**b**) of the bulk particle

Microstructural Changes

Control Sample (Before Exposure)

A control sample taken from the as-received batch and subjected to no gas exposure showed uniform elemental distribution within inner grains with roughly equal concentrations of Fe and Cu, based on EDX analysis. The location of a grain within the bulk particle did not affect these results.

Sample Quenched After the First Oxidation

The composition of the sample exposed to air for three hours showed negligible microstructural variations from the control sample in the spinel grains (Figs. 4a,5a). No significant differences were observed between the center grain and the edge grain (Fig. 4a, b).

Sample Quenched After the Ox–Red Cycle

The sample that had been exposed to air followed by 10CO–90Ar showed ubiquitous microstructural alterations within inner grains (grain aggregates) making up the carrier particle. The spinel grains broke down to at least two separate phases upon reduction (Fig. 5b). EDX analysis indicated a correlation between the dark regions and Fe concentration as well as the light, needle-like regions and Cu concentration. Additionally, it was observed that Fe-rich micron particles were beginning to appear around the edges of the bulk particle during the reduction exposure. It is not certain if these particles are a new phase that formed or the original matrix depleted in

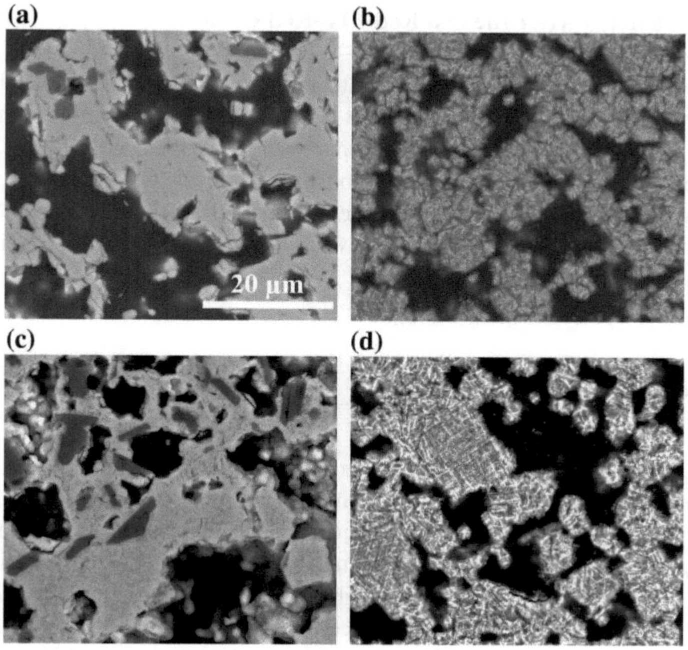

Fig. 5 Successive SEM images at 5000× of center spinel grains after Ox (**a**), Ox–Red (**b**), Ox–Red–Ox (**c**), and Ox–Red–Ox–Red (**d**)

Cu (or enriched with Fe). It was reported that at 800 °C in a reducing atmosphere of pure CH_4, the $CuO \cdot Fe_2O_3$ solid solution transformed into magnetite (Fe_3O_4), wüstite (FeO), delafossite ($CuFeO_2$), cuprous oxide (Cu_2O), and metallic Cu via lattice oxygen transfer and direct decomposition [11]. These phases may have existed in the present samples, which would explain the presence of the Fe-rich and Cu-rich regions. Further analysis using transmission electron microscopy is underway to confirm. Finally, grain boundaries on the order of a few hundred nanometers in width started to be visible at this stage.

Sample Quenched After the Ox–Red–Ox Cycle

The elemental distribution within the spinel phase after the Ox–Red–Ox gas exposure was found to be more uniform than in the quenched sample from the Ox–Red cycle. The Fe-rich particles on the surface of the bulk particle were still present. This is corroborated by the findings in Wang et al. that showed differentiation between the Fe-rich and Cu-rich phases after reoxidation [12]. A gradient in the oxygen partial pressure across the particle/grain aggregate may affect phase stability. Nanometer-scale voids formed within the matrix grains (Fig. 5c).

Sample Quenched After the Ox–Red–Ox–Red Cycle

This exposure series' results are somewhat similar to the Ox–Red sample. The phase separation within the grains was apparent (Fig. 5d). However, unlike in the O–Red exposure, the Cu-rich phase tended to form closer to the grain edges.

The Fe-rich formations on the surface observed after the Ox–Red cycle were also present in this sample. This time, they were found across the bulk particle, but not ubiquitous. Additionally, some of them were observed to be accompanied by an adjacent, but distinct, Cu-rich formation.

Conclusions

NETL's Cu/Fe spinel was investigated at 800 °C in both oxidizing and reducing atmospheres using a CSLM equipped with a furnace. Volume changes in individual particles were measured in situ to characterize how the material is altered by this high-temperature exposure. Particle volumes were noted to decrease during initial oxidation and then found to increase slightly during the subsequent reduction and oxidation exposures. SEM analysis showed that samples were relatively unchanged during initial oxidation but exhibited significant microstructural changes after the subsequent reduction, showing a separation of Cu- and Fe-rich regions, each with distinct morphology. The material also returned to a more uniform elemental distribution if oxidized a second time. However, the phase separation that was observed could result in changes in the oxygen carrier's performance over longer periods of time; future work is planned to investigate this question.

Acknowledgements This work was performed in support of the US Department of Energy's Fossil Energy Advanced Combustion Program. The research was executed through NETL Research and Innovation Center's Advanced Combustion effort. Research performed by AECOM Staff was conducted under the RES contract DE-FE-0004000. The authors wish to thank Mr. Matt Fortner for metallography.

Figure 1 reprinted with permission from Nealley et al.: Springer, JOM, Structural changes and material transport in Al_2O_3-supported Cu/Fe spinel particles in a simulated chemical looping combustion environment, Nealley WHH, Nakano A, Nakano J, Bennett JP, Copyright (2018).

The authors declare that they have no competing financial interest.

References

1. Metz B, Davidson OR, Bosch PR, Dave RL, Meyer A (eds) (2007) Mitigation. Contribution of working group III to the fourth assessment report of the intergovernmental panel on climate change, IPCC, 2007. Cambridge University Press, Cambridge
2. Hossain MM, de Lasa HI (2008) Chemical-looping combustions (CLC) for inherent CO_2 separations—a review. Chem Eng Sci 63:4433–4451
3. Nealley WHH, Nakano A, Nakano J, Bennett JP (2018) Structural changes and material transport in Al_2O_3-supported $CuFe_2O_4$ particles in a simulated chemical looping combustion environment. JOM 70(7):1232–1238
4. Lyngfelt A (2011) Oxygen carriers for chemical looping combustion—4000 h of operational experience. Oil Gas Sci Tech 66:161–172
5. Wang X, Chen Z, Hu M, Tian Y, Jin X, Ma S, Xu T, Hu Z, Liu S, Guo D, Xiao B (2017) Chemical looping combustion of biomass using metal ferrites as oxygen carriers. Chem Eng J 312:252–262
6. Voitic G, Hacker V (2016) Recent advancements in chemical looping water splitting for the production of hydrogen. RSC Adv 6:9867–9896
7. Hu W, Donat F, Scott SA, Dennis JS (2016) The interaction between CuO and Al_2O_3 and the reactivity of copper aluminates below 1000 °C and their implication on the use of the Cu–Al–O system for oxygen storage and production. RSC Adv 6:113016–113024
8. Liu W, Ismail M, Dunstan MT, Hu W, Zhang Z, Fennell PS, Scott SA, Dennis JS (2015) Inhibiting the interaction between FeO and Al_2O_3 during chemical looping production of hydrogen. RSC Adv 5:1759–1771
9. Bayham S, Straub D, Weber J (2017) Operation of the NETL chemical looping reactor with natural gas and a novel Copper-Iron material. NETL-PUB-20912; NETL technical report series; U.S. Department of Energy, National Energy Technology Laboratory, Morgantown, WV
10. Siriwardane R, Riley J, Bayham S, Straub D, Tian H, Weber J, Richards G (2018) 50-kWth methane/air chemical looping combustions tests with commercially prepared CuO–Fe_2O_3-alumina oxygen carrier with two different techniques. App Energy 213:92–99
11. Wang B, Yan R, Zhao H, Zheng Y, Liu Z, Zheng C (2011) Investigation of chemical looping combustion of coal with $CuFe_2O_4$ oxygen carrier. Energy Fuels 25:3344–3354
12. Wang B, Ma Q, Wang W, Zhang C, Mei D, Zhao H, Zheng C (2017) Effect of reaction temperature on the chemical looping combustions of coal with $CuFe_2O_4$ combined oxygen carrier. Energy Fuels 31:5233–5245

Part III
Thermodynamic and Mechanical Properties

Surface Tension of High Temperature Liquids Evaluation with a Thermal Imaging Furnace

Mindy Wu, Andrew H. Caldwell and Antoine Allanore

Abstract At high temperature, the reactivity of liquid metals, salts, oxides, etc. often requires a container-less approach for studying composition-sensitive thermodynamic properties, such as component activities and surface tension. This experimental challenge limits access to essential properties, and therefore our understanding of molten systems. Herein, a thermal imaging furnace (TIF) is investigated as a mean of container-less study of molten materials via the formation of pendant drops. In situ optical characterization of a liquid metal drop is proposed through the use of a conventional digital camera. We report one possible method for measuring surface tension of molten systems using this pendant drop technique in conjunction with an image analysis procedure. Liquid copper was used to evaluate the efficacy of this method. The surface tension of liquid copper was calculated to be 1.159 ± 0.043 Nm^{-1} at 1084 ± 20 °C, in agreement with published values.

Keywords Liquid copper · High temperature · Surface tension
Melt · Container-less

Introduction

Knowledge of the basic thermodynamic information and physical properties of the liquid state of materials, such as molten metals or alloys, molten oxide and molten intermetallic compounds, is relatively poorly developed, when compared with the breadth of data available for materials systems near room temperature. These data include, and are not limited to, the Gibbs energy, component activities, surface tension, melting temperature, and vapor pressure. The absence of this sort of information constrains the use and study of materials, not only for molten systems but also for the solid state, since many of these materials are employed at temperatures in the vicinity of T_{fus}.

M. Wu · A. H. Caldwell · A. Allanore (✉)
Department of Materials Science and Engineering, Massachusetts Institute of Technology,
77 Massachusetts Avenue, Cambridge, MA 02139, USA
e-mail: allanore@mit.edu

J. Nakano et al. (eds.), *Advanced Real Time Imaging II*, The Minerals,
Metals & Materials Series, https://doi.org/10.1007/978-3-030-06143-2_4

This lack of high temperature liquid state data can be attributed to the experimental challenges that accompany measurements of melting temperature, surface tension, etc. Of primary concern is the reactivity of molten materials at high temperature, as small amounts of impurities can significantly change the measured values of the quantities stated above. The issue of sample contamination from a holder or stage may be addressed by adopting a container-less approach to the study of molten systems. The pendant drop technique is one such method. For this technique, a rod of the material is suspended in a furnace, and localized melting at one end is induced to form a stable liquid drop. For example, pendant drops were recently used to measure Gibbs energy in molten alumina at temperatures in excess of 2000 °C [1].

The pendant drop technique is particularly conducive to the study of liquid surface tension due to the known dependence of the equilibrium drop shape on the two forces acting on the drop: gravity and surface tension. Previous studies have used the pendant drop technique to measure the surface tension of liquid metals [2–4] and refractory oxides [5–7]. With regards to the experimental setup for liquid metals at high temperature, the approaches to date have employed either refractory oxide capillaries to contain the liquid metal, which are unsuitable for high melting temperature metals, or electron beam or induction melting, which can only be used with electronically conductive liquids and may induce unwanted vibrations [8]. Both issues may be addressed by using a thermal imaging furnace (TIF) with an optical heating source.

In this paper we describe a method for measuring surface tension using a TIF in conjunction with an image analysis procedure that employs an open-source drop shape optimization program (OpenDrop [9]). To our knowledge, this is one of the first studies reporting an experimental setup suitable for surface tension measurements from container-less pendant drops of both liquid metal and nonmetal compounds at high temperatures. Copper was chosen as a model test liquid for evaluating the TIF measurement apparatus, as its surface tension is well-studied, but the approach may be used for higher melting temperature metals and compounds as well. Herein, the results for liquid Cu are presented.

Background

The relationship between the drop shape and surface tension (γ) is defined mathematically by the mean curvature of the drop

$$\frac{\mathrm{d}\phi}{\mathrm{d}s} + \frac{\sin\phi}{x} = \frac{\Delta P}{\gamma} = \frac{\Delta\rho g\,(h - z)}{\gamma}. \tag{1}$$

Equation (1) is the Young-Laplace equation in cylindrical coordinates (see Fig. 1) where ϕ is the tangent angle and s is the arc length with origin at $x = 0$. ΔP, the pressure difference across the drop interface, may be expressed as $\Delta\rho g\,(h - z)$ where

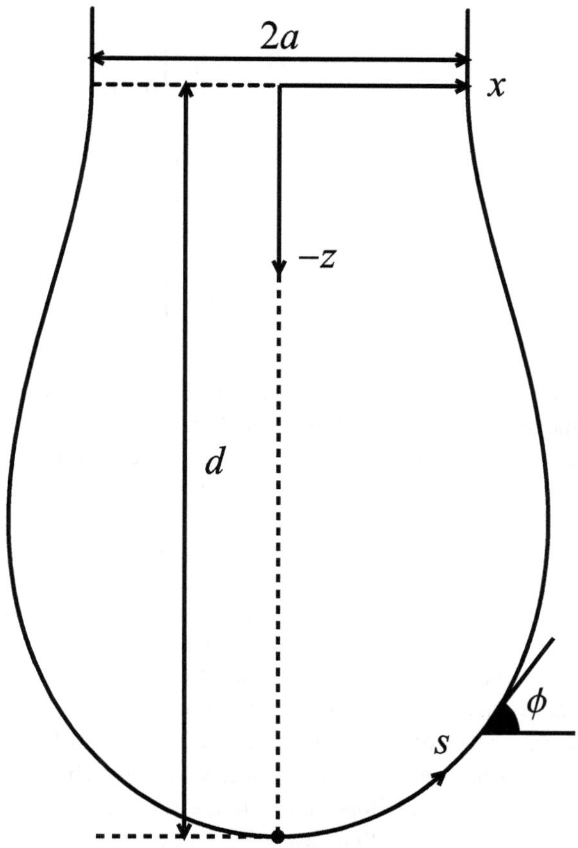

Fig. 1 Coordinate system for defining the shape of pendant drops

$\Delta\rho$ is the difference in density between the drop and the surrounding medium, g is the gravitational acceleration, and $z = h$ defines the height at which $\Delta P = 0$.

Equation (1) may be nondimensionalized by normalizing the coordinates with respect to the capillary length $L_c = \sqrt{\gamma/\Delta\rho g}$ (e.g., $\overline{x} = x/L_c$), yielding

$$\frac{\mathrm{d}\phi}{\mathrm{d}\overline{s}} = \left(\overline{h} - \overline{z}\right) - \frac{\sin\theta}{\overline{x}}. \tag{2}$$

The complete set of equations describing the family of pendant drop shapes is then given by Eq. (2) and Eqs. (3–7)

$$\frac{\mathrm{d}\overline{x}}{\mathrm{d}\overline{s}} = \cos\phi\,(\overline{s}) \tag{3}$$

$$\frac{\mathrm{d}\overline{z}}{\mathrm{d}\overline{s}} = \sin\phi\,(\overline{s}) \tag{4}$$

$$\bar{x} = \begin{cases} 0 & \bar{s} = 0 \\ \bar{a} & \bar{s} = \bar{s}_0 \end{cases} \tag{5}$$

$$\bar{z} = \begin{cases} -\bar{d} & \bar{s} = 0 \\ 0 & \bar{s} = \bar{s}_0 \end{cases} \tag{6}$$

$$\phi = \begin{cases} 0 & \bar{s} = 0 \\ \chi & \bar{s} = \bar{s}_0 \end{cases} \tag{7}$$

where $\bar{a}$ is the nondimensionalized rod radius, $\bar{d}$ is the nondimensionalized depth of the drop, $\bar{s}_0$ is the nondimensionalized arc length at $\bar{z} = 0$, and χ is the tangent angle at the solid-liquid interface ($\bar{s} = \bar{s}_0$). A numerical solution of the drop shape ($\bar{x}(s)$, $\bar{z}(s)$) may be obtained by renormalizing the domain with respect to $\bar{s}_0$.

The Bond number

$$Bo = \Delta\rho g R_0^2/\gamma = (R_0/L_c)^2 \tag{8}$$

is a dimensionless quantity that is a measure of the relative influence of gravity to that of surface tension on the drop shape [10, 11]. R_0 is the radius of curvature at $s = 0$. For an axisymmetric drop (this analysis is only true for axisymmetric drops) $R_0 = R_1(0) = R_2(0)$ where $R_1(s)$ and $R_2(s)$ are the principal radii of curvature. Large values of Bo are associated with drops of low sphericity. Small values of Bo indicate that the drop shape is largely controlled by surface tension. As a consequence, drops with small Bond numbers are relatively insensitive to d (drop growth). The error associated with measured values of γ is therefore much greater for drops with small Bo, as most drops will exhibit only small deviations from sphericity, which are difficult to capture with precision. Berry et al. [9] provides a detailed analysis of this phenomenon. Practically speaking, it is advantageous, therefore, to carry out experiments in which the pendant drops that form have Bond numbers that are as large as possible. Figure 2 provides some sense of the relation between rod diameter, drop size, and Bond number. Plotted are curves of Bo as a function of $\bar{d}$ for a range of nondimensionalized rod radii, $\bar{a}$, from 0.2 to 1.5. The dashed line is the locus of maximum drop volume. For perspective, we may look at typical liquid metals, which have capillary lengths near 5 mm. For a 2.5 mm radius rod (a typical rod size for use in the TIF), the Bond number for a liquid metal drop close to its maximum volume is approximately 0.45. From the analysis by Berry et al., it can be concluded that such a measurement, if performed, would yield values of the surface tension without significant error when computed via a curve fitting optimization. This of course does not take into account other sources of experimental uncertainty, such as sample contamination and vibrations.

The calculation of γ from pendant drop shapes was first conducted by Andreas et al. [12] and later expanded upon by Fordham [13]. In this approach, a shape factor $S = r_s/r_e$ is determined for the drop image, where r_e is the drop radius at its widest

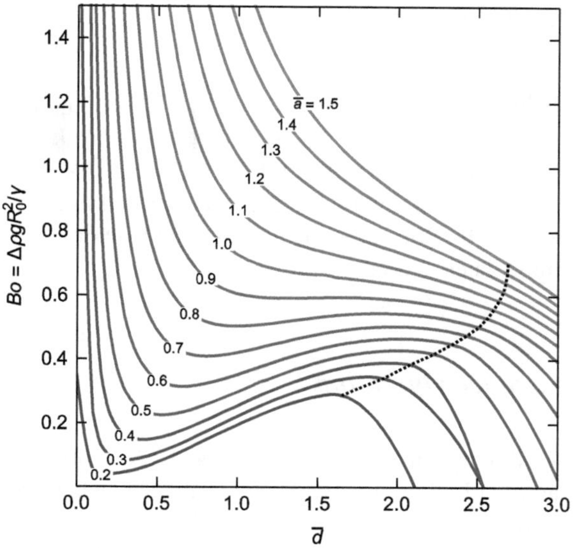

Fig. 2 Bond number (Bo) as a function of the nondimensionalized drop depth ($\overline{d} = d/L_c$) for a range of nondimensionalized rod radii ($\overline{a} = a/L_c$). The dashed line is the locus of maximum drop volume. The maximum drop volume is a point of unstable equilibrium, and values of Bo and $\overline{d}$ beyond this point to do not represent physically-realizable static drop shapes

point and r_s is the drop radius at $z = 2r_e - d$. From the Bond number equation ($Bo = \Delta\rho g R_0^2/\gamma$) a term $H(S) = 4Bo(r_e/R_0)^2$ is defined, and values of γ may be calculated from reported tables of H as a function of S. The principal deficiency of this approach is the relatively large error incurred by the high sensitivity of the calculated γ to the single quantity r_e. For a 1 % error in r_e, the uncertainty in γ may be as large as 20 % [14]. Nevertheless, the approach of Andreas and Fordham remains a standard analytical method for calculating surface tension from pendant drop images.

Greater precision in the determination of γ was achieved by the development of computer optimization programs for fitting the entirety of the drop profile to the theoretical drop shape [9, 15, 16]. In this method, the surface tension is calculated by minimizing an objective function of the unknown parameters governing the drop shape:

$$\underset{\nu}{\operatorname{argmin}} = \sum_{n=1}^{N} \left[e\left(u_n, \nu\right)\right]^2, \ \nu \equiv f\left(Bo, R_0, x(s_0), z(s_0), \ldots\right) \tag{9}$$

where $e\left(u_n, \nu\right)$ is the distance between point u_n on the measured drop profile and the nearest point on the theoretical drop curve ν, which is a function of several unknown parameters, invariably including the Bond number, R_0, and the position of the drop apex. For a detailed perspective, see Rotenburg et al. [16] and Berry et al. [9].

Experimental Methods

Sample Preparation

Copper rods (Goodfellow 99.95 % O.F.H.C., ∅ = 5 mm, 100 mm length) were degreased with anhydrous acetone and placed in a 3 M nitric acid bath at room temperature for 30 s before being rinsed with deionized water. They were then dried and immediately stored in Ar.

Furnace Operation

A TIF (Crystal Systems Corp., model TX-12000-I-MIT-PC) equipped with four Xe lamps (12 kW total) was used to form the pendant drops, a schematic of which is presented in Fig. 3. The metal rod was held in place by a stainless steel set-screw sample holder fixed to the upper shaft, which provides rotation and z-axis displacement via external stepper motors with sub-millimeter precision. In this way, deviations from drop axisymmetry were minimized. Sealing for the upper shaft was provided by an Ultra-Torr fitting. A quartz tube (Technical Glass Products, ∅ = 50 mm) sealed with Viton O-rings provided a gas-tight environment for the rod. The quartz tube was manufactured with a 23 mm flat circular pane positioned at the z-position of the furnace hot-zone and was oriented to face the welding glass viewing port on the furnace enclosure through which images were taken. The images were captured

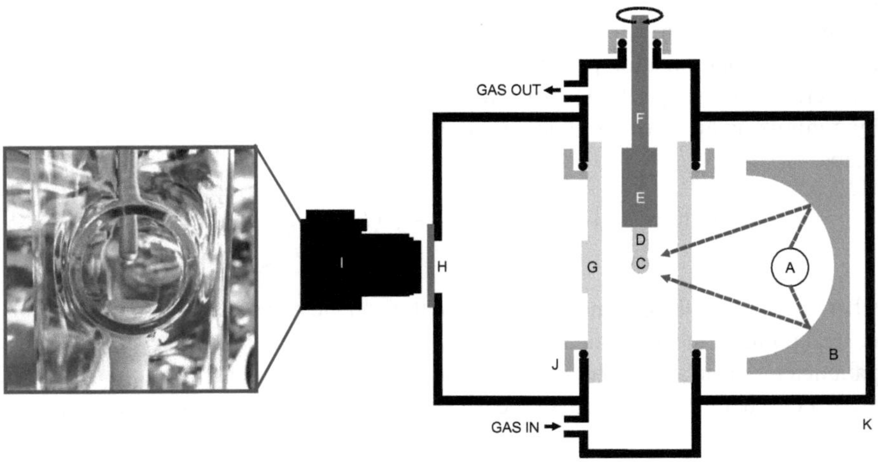

Fig. 3 **a** Xenon lamp. **b** Ellipsoidal mirror. **c** Liquid drop. **d** Sample rod. **e** Stainless steel sample holder. **f** stainless steel rod. **g** Quartz tube with window. **h** Welding glass. **i** Camera. **j** Threaded steel cap. **k** TIF

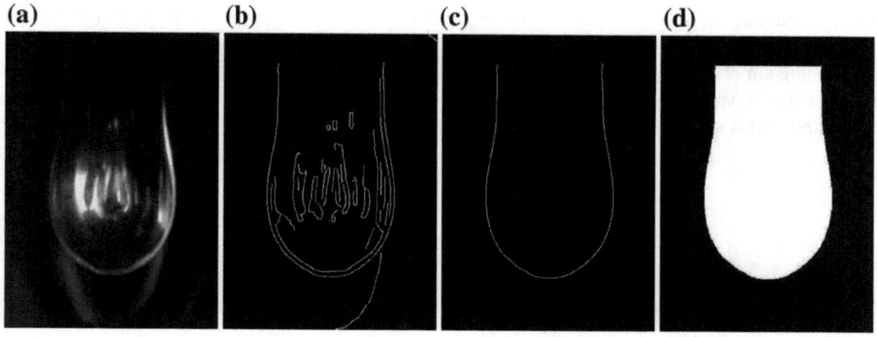

Fig. 4 **a** Raw drop image, cropped from the original photograph. **b** Image after passing through the Mathematica edge detection function. **c** The drop profile after removal of extraneous edges. **d** Final drop profile to be used with the OpenDrop program

using a tripod-mounted camera (Canon Inc., EOS Rebel T5i DSLR) equipped with a zoom lens (Canon Inc., EF-S 18–135 mm).

Prior to heating, the interior of the quartz tube was evacuated and refilled with Ar gas (Airgas Inc., UHP Ar > 99.999 %). This purge operation was carried out a minimum of three times, after which an Ar flow rate of 200 mL min^{-1} was established using a digital mass flow controller (Tylan General Inc., FC-260V). The lamps were then powered and the rotating (3 rpm) rod positioned so that its free end was located in the furnace hot-zone. The lamp power was increased by 1 % per minute until melting was observed. The temperature was measured with a type C thermocouple in the vicinity of the drop, with the uncertainty being within 50 °C of T_{fus} = 1084 °C. At this point rotation was stopped. Once the liquid copper drop was stable and of sufficient size, photographs were taken with the following image capture settings: 400 ISO, 1/30 s shutter speed, and f/5.6. An example of such an image is presented in Fig. 4a. If, at some point the drop detached from the rod, the rod end was refrozen and the lamp power ramp-up was carried out again.

Image Processing

The raw images were processed using Wolfram Mathematica 11. After cropping the images, an edge-detection algorithm was applied to isolate the drop profile as shown in Fig. 4b and c. The drop outline was then filled white (Fig. 4d). The processed images were subsequently analyzed with the open-source tensiometry software OpenDrop [9], which applies a drop shape optimization routine to calculate the apparent surface tension from the measured drop shape.

Fig. 5 Comparison of published values [17] for the surface tension of liquid copper with the value calculated in this study

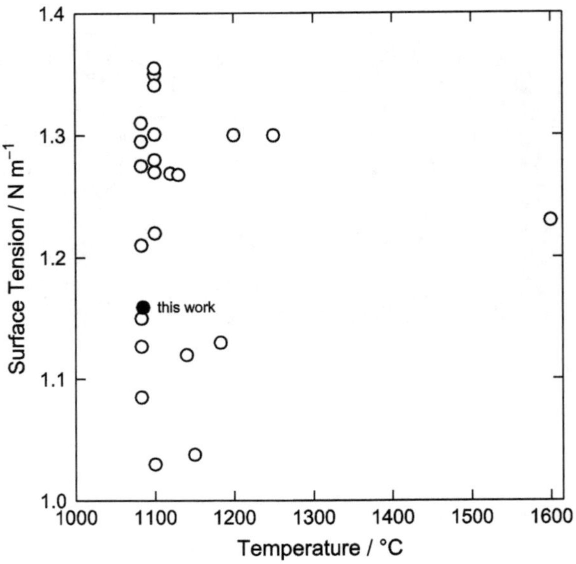

Results and Discussion

The calculated surface tension for liquid copper is $1.159 \pm 0.043\,\mathrm{Nm}^{-1}$ at a temperature of within 50 °C of T_{fus} (1084 °C). The error is the statistical uncertainty from 36 drop images. The value we calculate is in agreement with previously published values [17]. These are shown in Fig. 5. Our value is slightly lower than the typically accepted value of $1.3\,\mathrm{Nm}^{-1}$. The discrepancy may be due to surface active impurities such as oxygen, as some oxide was observed on the drop surface during the measurements.

Conclusion

A procedure for measuring surface tension from pendant drops using an optical TIF in conjunction with an open source tensiometry program OpenDrop was described. This is the first study employing a setup suitable for measuring the surface tension of both electronically conductive liquids and nonmetal liquids such as molten oxides. Liquid copper was chosen was a suitable candidate metal for evaluating this methodology. We report a value of $1.159 \pm 0.043\,\mathrm{Nm}^{-1}$ for the surface tension of liquid copper at its melting point, in agreement with published values.

Acknowledgements The authors are grateful to Prof. Osamu Takeda (Tohoku University) and Ms. Melody Wang (MIT) for their pioneering contributions with TIF furnace in our laboratory. Support for Mr. Andrew Caldwell comes from National Science Foundation (NSF), under grant number 1562545.

References

1. Nakanishi BR, Allanore A (2017) J Electrochem Soc 164(13):E460–E471
2. Allen BC (1963) Trans Metall Soc AIME 227:1175–1183
3. Man KF (2000) Int J Thermophys 21(3):793–804
4. Ricci E, Giuranno D, Sobczak N (2013) J Mater Eng Perform 22(11):3381–3388
5. Kingery WD (1959) J Am Ceram Soc 42(1):6–10
6. Rasmussen JJ, Nelson RP (1971) J Am Ceram Soc 54(8):398–401
7. Lihrmann JM, Haggerty JS (1985) J Am Ceram Soc 68(2):81–85
8. Peterson AW, Kedesdy H, Keck PH, Schwarz E (1958) J Appl Phys 29(2):213–216
9. Berry JD, Neeson MJ, Dagastine RR, Chan DY, Tabor RF (2015) J Colloid Interface Sci 454:226–237
10. Bashforth F, Adams JC (1883) An attempt to test the theories of capillary action: by comparing the theoretical and measured forms of drops of fluid. Cambridge University Press, Cambridge, UK
11. Merrington AC, Richardson EG (1947) Proc Phys Soc 59(1):1–13
12. Andreas JM, Hauser EA, Tucker WB (1937) J Phys Chem 42(8):1001–1019
13. Fordham S (1948) Proceedings of the royal society A: mathematical. Phys Eng Sci 194(1036):1–16
14. Stauffer CE (1965) J Phys Chem 69(6):1933–1938
15. Maze C, Burnet G (1969) Surf Sci 13(2):451–470
16. Rotenberg Y, Boruvka L, Neumann AW (1983) J Colloid Interface Sci 93(1):169–183
17. Allen BC (1972) In: Beer SZ (ed) Liquid metals: chemistry and physics. Marcel Dekker Inc., New York, pp 161–212

New Laue Micro-diffraction Setup for Real-Time In Situ Microstructural Characterization of Materials Under External Stress

D. Popov, S. Sinogeikin, C. Park, E. Rod, J. Smith, R. Ferry, C. Kenney-Benson, N. Velisavljevic and G. Shen

Abstract Laue X-ray diffraction (XRD) is a powerful probe to characterize pressure-/strain-induced microstructural changes in materials. The use of brilliant synchrotron radiation allows Laue XRD to be measured in a fast manner, leading to microstructural characterization, such as two-dimensional maps of single-crystals, their texture, and deformation, to be made in time-resolved mode with temporal resolution down to seconds. This technique can be very efficient in the studies of mechanisms of deformation, grain growth, recrystallization, and phase transitions. A progress has been obtained to extend application of Laue diffraction to high-pressure area. Recent case studies of $\alpha \rightarrow \beta$ transition in Si and $\alpha \rightarrow \omega$ transitions in Zr are briefly reported. A new experimental setup specifically optimized for real-time in situ Laue measurements has been developed at the 16-BMB beamline at the Advanced Photon Source. Due to the large X-ray energy range, which is typically up to 70 keV, a polychromatic beam diffraction technique can be efficiently implemented despite some limits introduced to the scattering angle by strain generation devices. Currently, the X-ray beam is focused at the sample position down to ~$2.2 \times 2.2\ \mu m^2$ at the full width at the half maximum. Precision sample translation stages provide fast data collection with step sizes down to 0.5 μm.

Keywords Laue diffraction · Microstructure · High pressure
Synchrotron radiation

D. Popov (✉) · C. Park · E. Rod · J. Smith · R. Ferry · C. Kenney-Benson · G. Shen
X-Ray Science Division, Argonne National Laboratory, HP-CAT, Argonne,
IL 60439, USA
e-mail: dpopov@anl.gov

S. Sinogeikin
DAC Tools LLC, Naperville, IL 60565, USA

N. Velisavljevic
Shock and Detonation Physics Group, Los Alamos National Laboratory, Los Alamos,
NM 87545, USA

J. Nakano et al. (eds.), *Advanced Real Time Imaging II*, The Minerals, Metals & Materials Series, https://doi.org/10.1007/978-3-030-06143-2_5

Introduction

Laue micro-diffraction using synchrotron radiation is an efficient tool to investigate mechanical behavior of materials in situ under external stress [1–3]. An exciting advantage of Laue diffraction, compared to the widely implemented diffraction with monochromatic beam, is that the single-crystal diffraction data can be collected several orders of magnitude faster, which enables fast microstructural mapping of materials with time resolution down to seconds. Changes in crystal morphology and orientation, twinning, strain, dislocations, and other defects can be observed, even if they are essentially controlled by kinetics and take place only in limited periods of time. At high-pressure collaborative access team (HP-CAT), considerable effort has been undertaken to implement this powerful technique to investigate mechanisms of materials transformation under quasi-hydrostatic high-pressure conditions. For instance, an irreversible $\alpha \rightarrow \beta$ transition in Si under hydrostatic compression in a diamond anvil cell (DAC) was characterized by D. Popov, C. Park, C. Kenney-Benson, and G. Shen [3]. By doing real-time mapping across the sample, gradual changes in morphology of the parental single-crystals and growth of the product phase were observed on the time scale of minutes. The morphology of the coexisting α- and β-phases is presented schematically in Fig. 1. Areas of the high-pressure phase were elongated parallel to a <100> direction of the parental single-crystal indicating that nucleation and growth of the β-phase may have been controlled by defects. The single-crystals of α-Si exhibited measurable deformation. Qualitatively this deformation can be interpreted as 'twisting' about their longest dimensions by angles less than 0.2°.

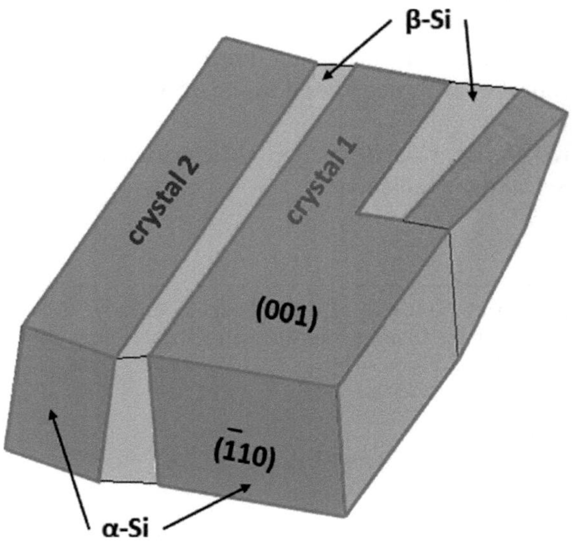

Fig. 1 Schematic of the appearance of the β-Si [3]. Crystals 2 and 3 are shown deformed assuming that they are twisted around their longest dimensions. Angles between the crystals are multiplied by a factor of 50

Grain enlargement is typically promoted by thermal energy to overcome activation barriers for diffusion, motion of dislocations, and migration of grain boundaries including newly formed ones. In some cases, however, pressure-induced phase transition results in grain enlargement in product phases at room temperature [4–7], and its cause may be fundamentally different from processes at high temperatures. Such pressure-induced grain enlargement has been reported in Zr based on comparison of X-ray diffraction patterns obtained with monochromatic beam on powdered α-Zr and on grainier ω-Zr [6, 7]. The first *in-operando* study of the grain enlargement process by real-time Laue diffraction mapping of the newly formed ω-Zr crystals has been conducted by D. Popov, N. Velisavljevic, C. Park, and G. Shen. The observed ω-Zr crystals nucleated locally, independently from each other, were growing gradually and continuously preserving their orientations during enlargement. The obtained results support a mechanism for the enlargement of the ω-Zr grains which are driven by energy stored during the deformation of the parental α-Zr. The widely known recrystallization process at high temperatures has the same driving force [8, 9]. The growth of ω-Zr crystals may be also driven by reduction of the total grain boundary area in a similar way as in the case of grain coarsening with 'abnormal' grain growth when selected grains grow up to much larger size than the average [8, 10].

Experimental Setup

A new experimental setup optimized for in situ high-pressure, time-resolved measurements using Laue micro-diffraction has been developed at the HP-CAT 16-BMB beamline of the Advanced Photon Source by the authors (Fig. 2). Different types of strain generation devices can be used in the setup, including deformation stages, indenters, other devices providing uniaxial compression/tensile, and DACs providing quasi-hydrostatic compression. For measurements with DAC, the setup is equipped with a membrane control system and a gearbox for remote tuning of pressure. A ruby fluorescence system is available on the side of the experimental table to adjust the DAC pressure before data collection.

All setup components are mounted on a granite table to maintain mechanical stability. A Perkin Elmer area detector (XRD1621) is used to record diffraction images. The angle of the area detector with respect to the incident X-ray beam can be changed in steps of 15° in the vertical plane using an adjustable detector arm, making it possible to collect Laue patterns in both transmitted and 90° geometries. The bending-magnet white X-ray beam is focused down to $2.2 \times 2.2\ \mu m^2$ at the sample position with KB mirrors. White X-rays with a 70 keV cutoff provide a reasonable number of reflections even in transmitted geometry, which is important since the 90° geometry is often difficult due to limitations on the scattering angle introduced by various strain generation devices. On demand, the incident flux can be increased by changing the focusing mirror angle (at a cost of lowering the upper X-ray energy cutoff).

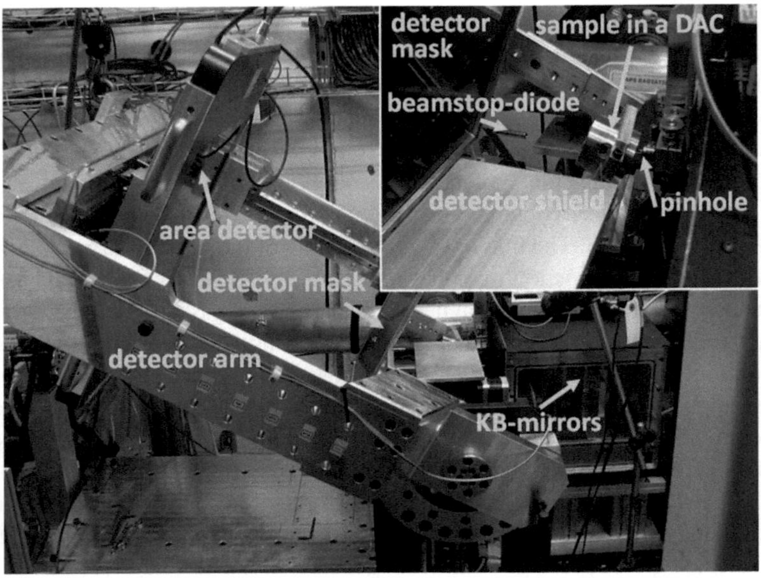

Fig. 2 Experimental setup at the 16-BMB beamline of Advanced Photon Source for in situ and in operando Laue micro-diffraction

Mutually perpendicular Newport horizontal translational stages and an Aerotech elevation stage with Newport XPS controllers are used for sample positioning. Samples mounted in strain devices with a mass up to 5 kg can be scanned in the two transverse dimensions with step sizes down to 0.5 μm. Multiple scans are collected in series in order to get sets of maps in real time. Strong reflections that can damage the area detector are blocked by a detector mask during data collection. A shield is introduced between the sample and area detector to protect the detector during the scanning process needed to align the sample to the X-ray beam. A Si111 channel-cut monochromator, switchable with the white beam in a matter of minutes, is used to calibrate detector geometry with the Dioptas [11] program, which is subsequently used for measurement of d-values of selected reflections and identification of powder-like phases.

Multiple data analysis programs are available at the beamline. Developed by D. Popov, the program PolyLaue, combined with Fit2d [12], can be used to identify single crystals by indexation of Laue reflections based on known unit cell parameters. The shape of the unit cell can be further refined based on the Laue data. In the case when multiple single crystals contribute to diffraction patterns simultaneously, orientation matrices of individual crystals are determined, and the predicted positions of diffraction spots are shown on diffraction images. PolyLaue can also generate maps of reflections. The XDI [13] program is also used to build maps of reflections but in a more efficient way by quantifying their intensities. The program LaueGo [14] maps lattice rotation and refines deviatoric strain. The current instrumentation allows determination of deviatoric strain down to ~ 10^{-4} level. LaueGo is also useful to calibrate the position and orientation of the area detector in 90° geometry.

Conclusions

Synchrotron radiation Laue diffraction is a powerful approach to investigate materials in situ and *in-operando* under external stress. Substantial progress in application of this technique to the high-pressure area has been recently obtained at HP-CAT. New experimental setup specifically optimized for in situ real-time measurements at high pressures has been developed at the HP-CAT 16-BMB beamline of Advanced Photon Source.

Acknowledgements We acknowledge the support of a DOE-BES grant under Award No. DEFG02-99ER45775 for this work, performed at HP-CAT (Sector 16), Advanced Photon Source (APS), Argonne National Laboratory. HP-CAT operations are supported by DOE-NNSA's Office of Experimental Sciences. The Advanced Photon Source is a US Department of Energy (DOE) Office of Science User Facility operated for the DOE Office of Science by Argonne National Laboratory under Contract No. DE-AC02-06CH11357. Los Alamos National Laboratory (LANL) is operated by LANS, LLC for the DOE-NNSA under Contract No. DE-AC52-06NA25396. The authors acknowledge funding support from LANL Science Programs 1 and 2.

References

1. Cornelius TW, Thomas O (2018) Progress of in situ synchrotron X-ray diffraction studies on the mechanical behavior of materials at small scales. Prog Mater Sci 94:384–434
2. Robach O, Kirchlechner C, Micha JS, Ulrich O, Biquard X, Geaymond O et al (2014) Laue microdiffraction at the ESRF. In: Barabash R, Ice G (eds) Chapter in "Strain and dislocation gradients from diffraction". Imperial College Press, UK
3. Popov D, Park C, Kenney-Benson C, Shen G (2015) High pressure Laue diffraction and its application to study microstructural changes during the $\alpha \to \beta$ phase transition in Si. Rev Sci Instrum 86:072204
4. McMahon MI (2012) High-pressure crystallography. Top Curr Chem-Ser 315:69–110
5. McMahon MI, Degtyareva O, Hejny C, Nelmes RJ (2003) New results on old problems: the use of single-crystals in high pressure structural studies. High Press Res 23(3):289–299
6. Velisavljevic N, Chesnut GN, Stevens LL, Dattelbaum DM (2011) Effects of interstitial impurities on the high pressure martensitic α to ω structural transformation and grain growth in zirconium. J. Phys. Condens. Matter 23:125402
7. Dewaele A, André A, Occelli F, Mathon O, Pascarelli S, Irifune T, Loubeyre P (2016) The $\alpha \to \omega$ phase transformation in zirconium followed with ms-scale time resolved X-ray absorption spectroscopy. High Press Res 36(3):237–249
8. Doherty RD, Hughes DA, Humphreys FJ, Jonas JJ, Jensen DJ, Kassner ME, King WE, McNelley TR, McQueen HJ, Rollett AD (1997) Current issues in recrystallization: a review. Mater Sci Eng A 238:219–274
9. Doherty RD, Gottstein G, Hirsch J, Hutchinson WB, Lucke K, Nes E, Wilbrandt PJ (1988) Report of panel on recrystallization textures: mechanisms and experiments. In: Kallend JS, Gottstein G (eds) ICOTOM 8, TMS

10. Novikov V (1997) Grain growth and control of microstructure and texture in polycrystalline materials. CRC Press, Boca Raton
11. Prescher C, Prakapenka V (2015) DIOPTAS: a program for reduction of two-dimensional X-ray diffraction data and data exploration. High Press Res 35(3):223–230
12. http://sector33.xray.aps.anl.gov/~tischler/
13. Hammersley AP (2016) FIT2D: a multi-purpose data reduction, analysis and visualization program. J Appl Crystallogr 49:646–652
14. Hrubiak R, Smith JS, Shen G (2017). XRD contrast imaging using hardware-based 'fly' scanning and XDI software at HPCAT. In: Poster presented at the 26th AIRAPT international conference on high pressure science and technology, Beijing, China, August 2017

Part IV
Iron & Steelmaking II

In Situ Study on the Transformation Behavior of Ti-Bearing Slags in the Oxidation Atmosphere

Yongqi Sun and Zuotai Zhang

Abstract Rutile acts as a target phase for the titanium (Ti) recovery from Ti-bearing blast furnace slags (Ti-BFS) due to its special properties. In this study, using single hot thermocouple technique (SHTT), we investigated the crystallization behaviors of the Ti-BFS and the target rutile precipitation behaviors where both the holding temperature, the basicity (mass ratio of CaO to SiO_2) and the P_2O_5 content were considered. We found that basicity has a vital influence on the crystallization behaviors and rod-like rutile only formed with lower basicity. As the basicity increased, the primary phase would transform from rutile to perovskite. As the basicity was 0.5, with the temperature increasing, the growth rate of rutile length initially increased, followed by a decrease with further increasing holding temperature. Thus, the growth rate of rutile had a maximum value of 7.74 μm/s at 1260 °C. Furthermore, the rutile growth followed a one-dimensional template, and the P_2O_5 content had an important impact. By increasing the content of P_2O_5, the incubation time of the rod-like rutile got decreased, suggesting that the rutile precipitation got much easier.

Keywords Ti-bearing slags · In situ study · Crystallization behaviors · Rutile Chemical modification

Introduction

China has the world's largest titanium resource, and more than 90% of them is located in southwestern China [1], existing in the form of vanadium titanium-magnetite ores. Generally, through blast furnace process, the valuable iron and vanadium elements are extracted. As a result, the titanium element remains in the blast furnace slags, i.e., titanium-bearing blast furnace slags (Ti-BFS). The content of TiO_2 in these Ti-BFS

Y. Sun (✉)
School of Chemical Engineering, The University of Queensland,
Brisbane, St Lucia, QLD 4072, Australia
e-mail: uqysun16@uq.edu.au

Z. Zhang
School of Environmental Science and Engineering, Southern University of Science and Technology, Shenzhen 518055, People's Republic of China

J. Nakano et al. (eds.), *Advanced Real Time Imaging II*, The Minerals,
Metals & Materials Series, https://doi.org/10.1007/978-3-030-06143-2_6

is around 22–25 wt% [2, 3], which accounts for a significant unutilized titanium-bearing resource. On the other hand, these Ti-BFS have not been effectively utilized, and as a result, more than 80 million tons (Mt) of Ti-BFS have been accumulated with a remaining large annual discharge rate [4]. Without timely treatment, these Ti-BFS will cause air and water pollution, land occupation and resource waste.

There are many methods developed for the treatment and utilization of Ti-BFS [5–10], and among them, selective crystallization and phase separation (SCPS) method has great advantages such as energy saving, high separation ratio and less contamination. Through, SCPS, the titanium element is enriched into the titanium-bearing mineral phase, and then the titanium-bearing phases are separated from other phases. The keys for the success of SCPS are the control of process condition such as the temperature and reaction atmosphere and modification of their inherent compositions and properties such as crystallization behaviors. The primary basicity (mass ratio of CaO to SiO_2) of Ti-BFS is ~1.2, and in this case, the precipitated titanium-bearing phase is $CaTiO_3$. Compared with $CaTiO_3$, rutile (TiO_2) has a good separation property with other titanium free phases due to the rod shape structure and its high density [5–7]. In order to transform the titanium-bearing phase from $CaTiO_3$ to rutile, the chemical compositions of the Ti-BFS should be modified. A reasonable clue is to restrict the link between CaO and TiO_2 and the precipitation of CaO from the Ti-BFS, and meanwhile enhance the precipitation of titanium element. Following this idea, some acidic additives, such as SiO_2 and P_2O_5, could be used to link with the CaO in the molten Ti-BFS, and as a result, the titanium elements got precipitated.

In this study, as a first step, we used SiO_2 to modify the chemical compositions of the Ti-BFS and decrease the basicity. Secondly, we used a small content of P_2O_5 to further modify the crystallization behaviors of the rutile. The selection of this two-step process could also decrease the cost of the SCPS method. Therefore, in the present study, we firstly investigated the effect of holding temperature on the crystallization behaviors of the Ti-BFS with the basicity of 0.5. Then, we investigated the effect of slag basicity with SiO_2 as a major chemical additive and the further effect of P_2O_5 working as a minor chemical additive.

Methodology

Sample Preparation

In this study, seven samples were prepared using analytically regent (AR) oxides of CaO, SiO_2, MgO, Al_2O_3, TiO_2 and P_2O_5, detailed in Table 1. It should be pointed out that the design of CSMAT-1, 2 and 3 mainly presents the effect of slag basicity, the design of samples CSMAT-1, 4 and 5 mainly presents the effect of TiO_2 content, while the design of CSMAT-1, 6 and 7 mainly shows the effect of P_2O_5 content.

Firstly, these oxides were weighed based on the chemical compositions in Table 1, fully mixed and placed in a molybdenum (Mo) crucible (Φ40 mm × 45 mm ×

Table 1 Chemical compositions of designed samples

No.	CaO	SiO_2	MgO	Al_2O_3	TiO_2	P_2O_5	Basicity
CSMAT-1	19	37	7	12	25	0	0.5
CSMAT-2	23	33	7	12	25	0	0.7
CSMAT-3	27	29	7	12	25	0	0.9
CSMAT-1	19	37	7	12	25	0	0.5
CSMAT-4	20	41	7	12	20	0	0.5
CSMAT-5	22	44	7	12	15	0	0.5
CSMAT-1	19	37	7	12	25	0	0.5
CSMAT-6	18	37	7	12	25	1	0.5
CSMAT-7	17	34	7	12	25	5	0.5

H40 mm). Secondly, the mixture was placed into the hot zone of the furnace and melted at 1550 °C under the high purity argon gas for 1 h to homogenize the chemical compositions, during which we also use a Mo wire to stir the molten slags to confirm its homogeneity. Thirdly, the liquid slags were rapidly poured into the cold water to obtain a glass sate. After that, the glassy slags were dried at 120 °C for 12 h, crushed and ground into 300 meshes for compositional analysis and the further crystallization experiments. The chemical compositions were measured by the X-ray fluoroscopy (XRF) technique, which shows small deviation with the designed ones. Furthermore, we also used X-ray diffraction (XRD) technique to confirm the glassy state of the prepared slags.

Crystallization Experiments

The crystallization behaviors of the Ti-bearing slags were characterized through a series of isothermal experiments using a single hot thermocouple technique (SHTT) [11, 12], which is schemed in Fig. 1. The images of the samples, with the resolution of 28.35 pixels/cm, were captured by a microscope equipped with a video camera and sent to a computer for analyzing the crystallization behaviors. This technique can heat/cool the samples and measure the samples simultaneously. Before experiments, the temperature was calibrated using pure K_2SO_4 sample with the melting point of 1067 °C. Firstly, around 10 mg glassy sample was mounted on the heating zone tip of the Pt–Rh thermocouple. Secondly, based on the temperature schedule, the samples were quickly heated to 1500 °C, fully melted and held for 120 s to decrease the gas bubbles, homogenize the temperature and chemical compositions. Thirdly, the liquid slags were quickly cooled to a preset temperature under the rate of ~ 60 K/s and held at this point for a long time (holding temperature, generally more than 30 min), during which the crystals would fully nucleate and grow in the molten slags. It should be pointed out that the isothermal experiments were overall in the range of 1360–1000 °C while it was quite different for each sample.

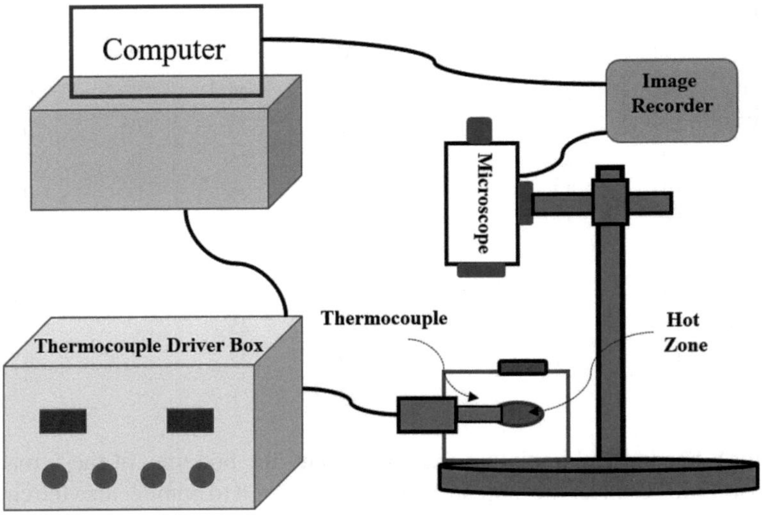

Fig. 1 Schematic diagram of the experimental system

As the temperature reaches the preset point, generally, it will take a timescale for the crystal to appear in the molten slags, which can be defined as incubation time, accounting for a key parameter to characterize the crystallization behaviors of slags. As the holding temperature varies, the incubation time will change correspondingly. Thus, we can use the time-temperature transformation (TTT) curve to indicate the relationship between the incubation time and the holding temperature. As the chemical compositions of the slags vary, the TTT curve will shift correspondingly, based on which the influences of chemical compositions, such as basicity and P_2O_5 content, can be clarified.

Results and Discussion

Effect of Holding Temperature

We firstly measured the crystallization behaviors of sample CSMAT-1 using the SHTT method. It was found that the holding temperature showed a significant influence on the crystalline precipitations in the molten slags. Regarding to the macro-crystallization phenomenon of the slags, here we presented two representative temperature points in Figs. 2 and 3. No post-image analysis was conducted to these figures where only the time period of holding experiments was presented. From Fig. 2, we can observe that at low temperatures (1040 °C), the undercooling degree was so large that the crystallization behavior could occur in all the areas of the molten slags including the center and the sides along the thermocouple. In this case,

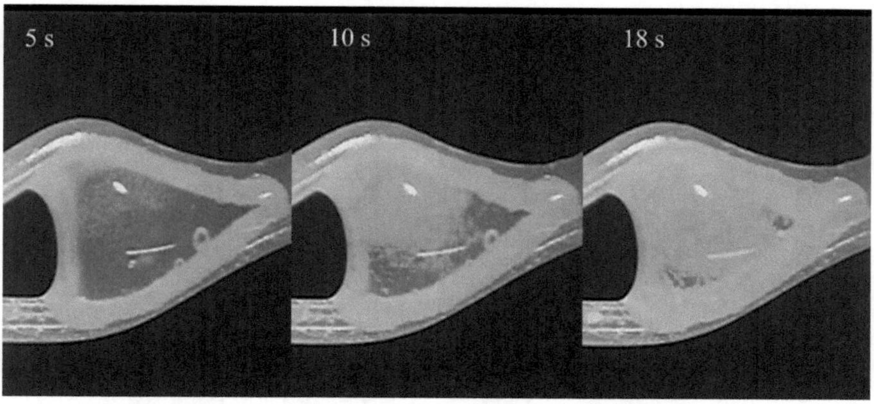

Fig. 2 Evolution of phases at low temperatures (1040 °C) in Ti-BFS

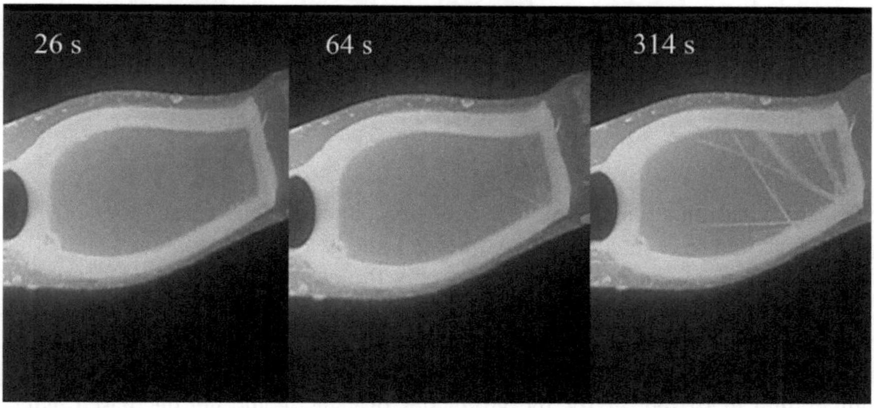

Fig. 3 Evolution of rod-like rutile at high temperatures (1240 °C) in Ti-BFS

homogeneous nucleation and crystal growth quickly took place due to the large undercooling degree. However, as the holding temperature increased, Fig. 3 presents that the nucleation and crystal growth only occurred in the side of the molten slags along the thermocouple. At high temperatures (1240 °C), the undercooling degree of the slags was so small that the nucleation only took place in the points along the thermocouple due to the local uneven surface, acting as the nucleation points. In other words, at high temperatures, heterogeneous nucleation and crystal growth occurred and only several rod-like rutile crystals, as confirmed in our previous study using X-ray diffraction [5], appeared, which was displayed in Fig. 3.

Based on the morphology of the rutile at high temperatures and the kinetic analysis, we can conclude that the rod-like rutile followed a one-dimension (1D) growth mode. Therefore, the growth rate of the rutile could be calculated based on the SHTT images. The diameter of the thermocouple was constant, namely, 30 um, which can be a reference of the rutile. Accordingly, the length of the rod-like rutile can be

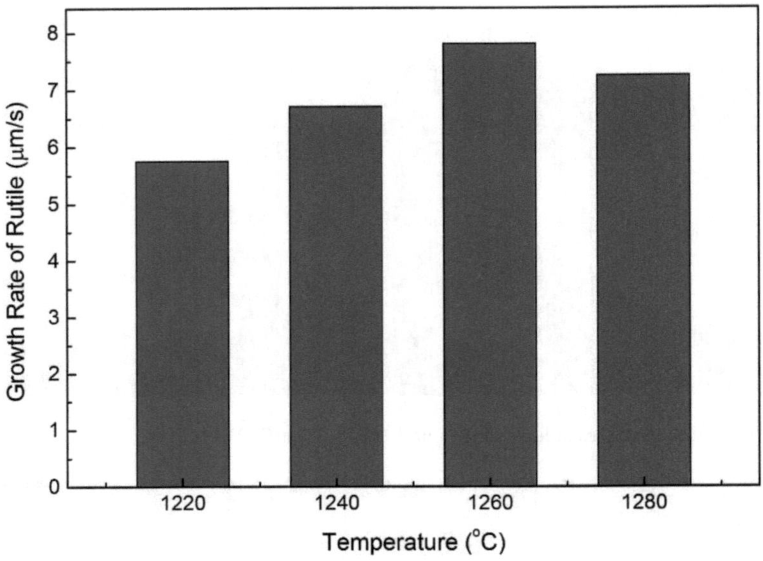

Fig. 4 Growth rate of rutile in sample CSMAT-1

calculated, and thus, its growth rate could be obtained. The results are shown in Fig. 4. It should be pointed out that both slag basicity and P_2O_5 content would affect the growth rate of the rutile and herein only the effect of holding temperature is displayed. As can be observed, with increasing holding temperature, the growth rate of the rod-like rutile first increased and then decreased, which could be analyzed based on the varying viscous behaviors and the undercooling degree. As the holding temperature increased, firstly, the viscosities of slags would get decreased, which would enhance the mass diffusion in the molten slags. Thus, the growth rate of the rutile firstly increased. As the holding temperature further increased, the viscosities of slags would further decrease, which would enhance the growth rate of rutile. On the other hand, with increasing holding temperature, the undercooling degree of the slags would get decreased, which would restrict the crystallization behaviors of the slags. The competence of these two effects finally resulted in a decreasing growth rate of the rod-like rutile with further increasing holding temperature.

Effect of Slag Basicity

Then we mainly analyzed the effect of basicity on the crystallization behaviors of the Ti-BFS, and herein three basicity values were considered, namely, 0.5, 0.7 and 0.9. The TTT curves of the samples are displayed in Fig. 5. As can be observed, the TTT curves were mainly composed two parts, namely, the high temperature range where Ti-bearing mineral phase acted as the only precipitated phase and the low

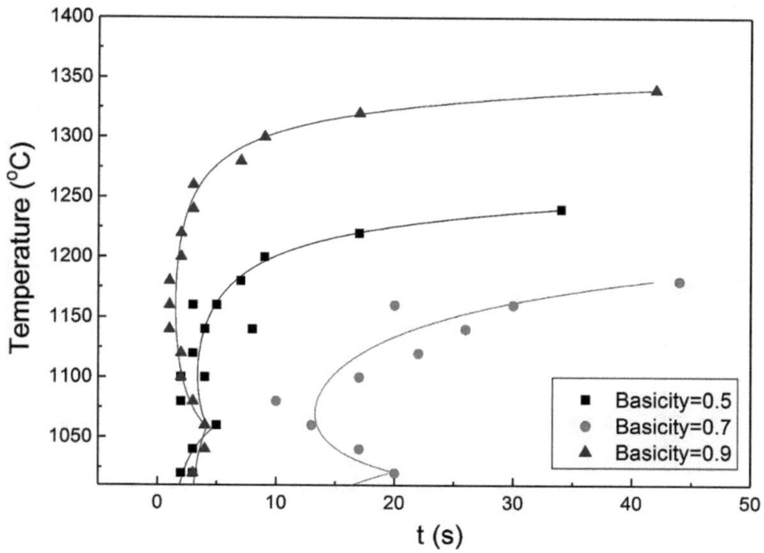

Fig. 5 TTT curves of the samples with varying basicities

temperature range where various minerals co-precipitated. In this study, the high temperature range was mainly considered where the Ti-bearing mineral phase was the only precipitated phase. Based on the high temperature range of the TTT curves in Fig. 5, several variation trends could be noted.

Firstly, the high temperature part of TTT curves where the rutile was precipitated followed a "C" shape, i.e., with increasing holding temperature, the incubation time of the rutile firstly decreased and then increased, indicating the crystallization ability of the Ti-BFS got firstly increased and then decreased by the increasing holding temperature. This was in agreement with the aforementioned growth rate of rutile length with increasing holding temperatures, which was also caused by the competence between the varying undercooling degree and viscosities with increasing holding temperature. Secondly, we can observe that with varying basicity, the TTT curves greatly shifted, i.e., the incubation time of the precipitated phase greatly changed. When the basicity was 0.9 and 0.7, the primary phase precipitated in slags was $CaTiO_3$ and when the basicity decreased to 0.5, the primary phase transformed to rod-like rutile. As the basicity decreased from 0.9 to 0.7, the decreasing CaO content would decrease the crystallization ability of $CaTiO_3$ from mass balance. Thus, the incubation time of $CaTiO_3$ increased. However, as the basicity decreased further from 0.7 to 0.5, the decreasing CaO content would enhance the crystallization ability of rutile. As a result, the incubation time further decreased.

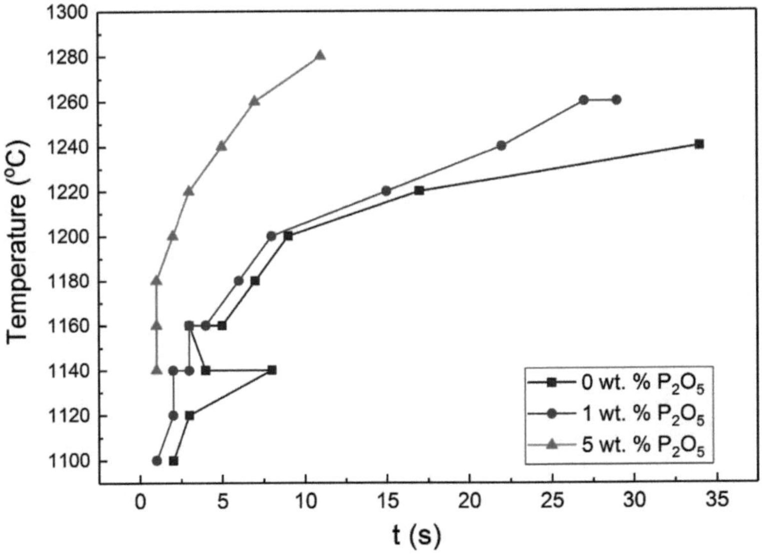

Fig. 6 TTT curves of the samples with varying P_2O_5 additions

Effect of P_2O_5 Content

As aforementioned, the modification of the basicity of Ti-BFS behaved as the first step, i.e., the major compositions of the Ti-BFS got modified using the low-cost material of SiO_2 and the basicity of the Ti-BFS has been modified to 0.5. Next, the chemical compositions of the Ti-BFS could be further optimized using other minor additives as a second step, and herein we chose P_2O_5, which was also a typical acidic oxide with a stronger acidity than SiO_2. The results are shown in Fig. 6. As can be observed, with increasing P_2O_5 content to 1 wt%, the incubation time of the rod-like rutile slightly decreased, indicating the crystallization ability of the primary phase got enhanced. With the P_2O_5 content further increasing to 5 wt%, the incubation time further decreased. Therefore, during the actual process, a reasonable P_2O_5 could be selected considering both the crystallization behavior optimizations and the process costs.

Conclusions

Titanium-bearing blast furnace slags (Ti-BFS) is an important titanium resource and our study provided fundamental evidence for the titanium extraction from the Ti-BFS using the selective crystallization and phase separation method. We found that as the basicity of slags was modified from 0.9 to 0.5, the primary phase precipitated in the Ti-BFS changed from $CaTiO_3$ to rod-like rutile, which had a better separation property. We also found that with increasing holding temperature, the crystallization ability of rutile first increased due to the decreasing slag viscosities and then decreased due to the decreasing undercooling degree of the Ti-BFS. Thus, the growth rate of the rutile presented a maximum value at 1260 °C. Next, we used P_2O_5 as a further additive and investigated the effect of P_2O_5 content. We found that as the P_2O_5 content increased, the incubation time of rutile got remarkably decreased, indicating the crystallization ability of the Ti-BFS got enhanced.

References

1. Wang LP et al (2004) Distribution and production status of titanium resources in China. Chin J Rare Metals 28:265–267
2. Wen L, Zhang JZ (2011) Properties on titanium-bearing blast furnace slags. J Iron Steel Res Int 23:1–3
3. Zhang L et al (2006) Effect of perovskite phase precipitation on viscosity of Ti-bearing blast furnace slag under the dynamic oxidation condition. J Non-Cryst Solids 352:123–129
4. Zhang L et al (2006) Dynamic oxidation of the Ti-bearing blast furnace slag. ISIJ Int 46:458–465
5. Li J, Zhang ZT, Wang XD (2011) Crystallization behavior of rutile in the synthesized Ti-bearing blast furnace slag using single hot thermocouple technique. ISIJ Int 51:1396–1402
6. Li J et al (2013) Influence of basicity and TiO_2 content on the precipitation behavior of the Ti-bearing blast furnace slags. ISIJ Int 53:1696–1703
7. Li J, Zhang ZT, Wang XD (2012) Precipitation behaviour of Ti enriched phase in Ti bearing slag. Ironmak Steelmak 39:414–418
8. Sun Y et al (2015) Co-modification and crystalline-control of Ti-bearing blast furnace slags. ISIJ Int 55:158–165
9. Sun Y et al (2014) The effect of P_2O_5 on the crystallization behaviors of Ti-bearing blast furnace slags using single hot thermocouple technique. Metall Mater Trans B 45:1446–1455
10. Li Z et al (2016) Effect of Al_2O_3 addition on the precipitated phase transformation in Ti-bearing blast furnace slags. Metall Mater Trans B 47:1390–1399
11. Kashiwaya Y et al (1998) Development of double and single hot thermocouple technique for in site observation and measurement of mold slag crystallization. ISIJ Int 38:348–356
12. Kashiwaya Y et al (1998) An investigation of the crystallization of a continuous casting mold slag using the single hot thermocouple technique. ISIJ Int 38:357–365

Dissolution of Sapphire and Alumina–Magnesia Particles in $CaO–SiO_2–Al_2O_3$ Liquid Slags

Hamed Abdeyazdan, Neslihan Dogan, Raymond J. Longbottom, M. Akbar Rhamdhani, Michael W. Chapman and Brian J. Monaghan

Abstract Understanding the dissolution kinetics of non-metallic inclusions in liquid slag is key in optimization of slag composition for inclusion removal. In this study, the rate of dissolution of high-precision spheres of sapphire and alumina–magnesia particles in $CaO–SiO_2–Al_2O_3$ liquid slags was measured in situ using a laser scanning confocal microscope at 1500 °C. It was found that the rate of dissolution of both sapphire and alumina–magnesia particles increased when the slag basicity is increased. A layer was observed around the dissolving sapphire. This layer may be a product layer and/or indicative of a mass transfer rate-controlling system. In the case of alumina–magnesia particle, the kinetics appeared more complex and depended on slag composition. No product layer or mass transfer layer was observed around the particle dissolving in slag with low basicity, whereas for the high basicity slag, a product or stagnant layer was observed, similar to that of the sapphire particle. Assuming a mass transfer-controlled system, measured diffusion coefficients for sapphire particles in slags tested in this study ranged from 10^{-11} to 10^{-10} $m^2 s^{-1}$ at 1500 °C.

Keywords Dissolution kinetics · Inclusion removal · Slag · Inclusions Clean steel

H. Abdeyazdan · R. J. Longbottom · B. J. Monaghan
PYROmetallurgical Research Group, School of Mechanical, Materials, Mechatronic and Biomedical Engineering, University of Wollongong, Wollongong, NSW 2522, Australia

M. Akbar Rhamdhani
Department of Mechanical and Product Design Engineering, Swinburne University of Technology, Hawthorn, VIC 3122, Australia

N. Dogan (✉)
Department of Materials Science and Engineering, McMaster University, Hamilton, ON L8S 4L7, Canada
e-mail: dogann@mcmaster.ca

M. W. Chapman
BlueScope Ltd, Port Kembla, NSW 2505, Australia

J. Nakano et al. (eds.), *Advanced Real Time Imaging II*, The Minerals, Metals & Materials Series, https://doi.org/10.1007/978-3-030-06143-2_7

Introduction

Non-metallic inclusions are harmful for steel productivity and physical properties [1, 2]. Inclusions are normally removed from liquid steel through reaction with the slag phase [2, 3]. In this process, inclusions must pass through the steel–slag interface to react with the slag [4–11]. This step is broadly dependant on the slag–inclusion phase wetting/interfacial tension and is favoured when the interfacial tension for the steel–inclusion is greater than interfacial tension for the slag–inclusion ($\sigma_{\text{steel–inclusion}} > \sigma_{\text{slag–inclusion}}$) [7–11]. After passing through the steel–slag interface, inclusions must dissolve in slag phase at a sufficiently fast rate. If dissolution is not sufficiently fast, there is a risk of inclusion re-entrainment [12]. Much research has been conducted to establish the effects of slag composition on dissolution rate. Such data are required if the dissolution mechanism is to be fully understood and the inclusion dissolution rate is to be optimized [12–45] with respect to slag composition. The high-temperature laser scanning confocal microscope (LSCM) provides in situ measurement of the dissolution rate of micro-particles. Kinetic studies to establish the rate-controlling step in the dissolution of inclusions using LSCM have mostly focused on typical oxide inclusions, Al_2O_3 [12, 30–41] and $MgAl_2O_4$ [33, 36, 40, 42] dissolving in $CaO–Al_2O_3–SiO_2–(MgO)$ slags. There are also a limited number of works on the dissolution of MgO, ZrO_2 and SiO_2 particles in slag [33, 36, 40, 43, 44]. While the use of the LSCM in inclusion dissolution studies has resulted a better understanding of the process, and more quantitative data for researchers and engineers use, there are still conflicts within the reported data that need to be resolved. In particular, there are conflicts on what are the rate-controlling mechanisms reported for the dissolution kinetics of typical steelmaking inclusions such as alumina and alumina–magnesia spinel in slag [12, 30–42]. Conflicting findings in the reported mechanisms using LSCM technique may in part be caused by the irregular shape of dissolving oxide particles used in these studies. Research by Michelic et al. [39] on the dissolution of alumina particles and Feichtinger et al. [44] for the dissolution of silica particles were the only studies that used high-precision spherical oxide particles. In all other studies, irregular particles have been used and were assumed to be spherical for the purpose of kinetic analyses. This would cause an error when the particle radius is measured. This error is a function of non-spheroidicity and how the particle radius would be apparently changing on rotation.

In this research, the dissolution of high-precision spheres of sapphire (alumina single crystal) and alumina–magnesia particles in $CaO–SiO_2–Al_2O_3$ liquid slags was studied using a LSCM. The change in the radius of particles with time was tracked. The dissolution rate-controlling mechanism was attempted to be established through not only analyses of the rate data, but also considering electro-optical characterization of the slag–inclusion reaction interface.

Table 1 Chemical composition of the slags used

Slag	Slag composition in mass%			$B\left(\frac{\%CaO}{\%SiO_2}\right)$
	CaO	SiO_2	Al_2O_3	
1	25.5	57.1	17.4	0.45
2	32.7	48.1	19.2	0.68
3	40.3	40.7	19.0	0.99

Experimental Methodology

Preparation of Slags and Single Crystal Spherical Particles

Dissolution experiments of sapphire (alumina single crystal) and alumina–magnesia oxide spheres were carried out using a laser scanning confocal microscope (LSCM) at 1500 °C with $CaO–SiO_2–Al_2O_3$ slags of compositions given in Table 1. Slags 1–3 were prepared by pre-melting appropriate mixtures of CaO, SiO_2 and Al_2O_3, quenching the fused slag in water and then crushing the resultant glass. This process was repeated to obtain a homogenous slag. The reported composition for slags 1–3 in Table 1 is based on XRF analysis. The synthetic inclusions were high-precision spheres obtained from Sandoz Fils SA. Their use is intended to minimize uncertainty in evaluating changes in size during dissolution.

Particle Dissolution Experiment

Thermal radiation often obscures the images of hot bodies. In LSCM, this issue is overcome using a pinhole that blocks any light that reaches the detector that does not emanate from the sample focal plane. In addition, the detector is turned to the LSCM light source, a He–Ne laser of wavelength 632.8 nm. This combination of pinhole and detector enables high-temperature imaging. Details of the LSCM technique are given elsewhere [32, 42].

Synthetic inclusions were added to the cold fused slag as shown in Fig. 1 which shows a schematic of the sample holder and crucible setup used in this study. The inclusion, sample holder and crucible were rapidly heated to the experimental temperature (1500 °C) in an infrared furnace and under an argon atmosphere. Once liquid, the slag was found to cover the inclusion particle, and the inclusion immersed in the slag and started to dissolve. This was defined as time zero. The particles of >0.2 mm dropped to the bottom of the crucible during the dissolution. This has been observed in previous studies [31, 33]. The end of the experiment was determined when the inclusion completely dissolved. Selected experiments

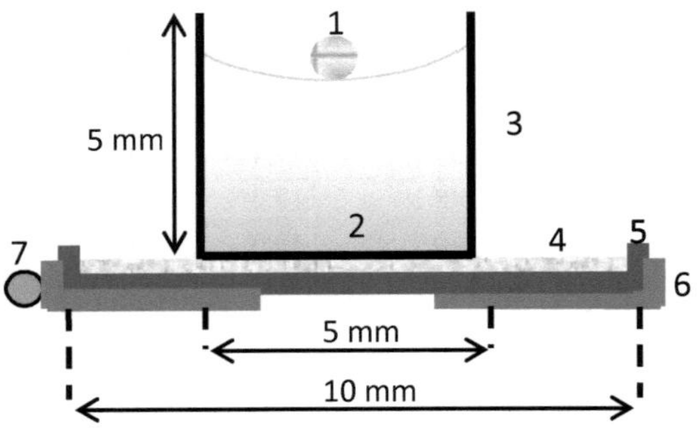

1: Sapphire/alumina-magnesia particle
2: Slag
3: Platinum crucible
4: Alumina powder
5: Aluminaboat
6: Crucible holder
7: LSCM thermocouple

Fig. 1 Schematic of sample and crucible configuration in LSCM

Fig. 2 Schematic of furnace heating profile applied in LSCM experiments

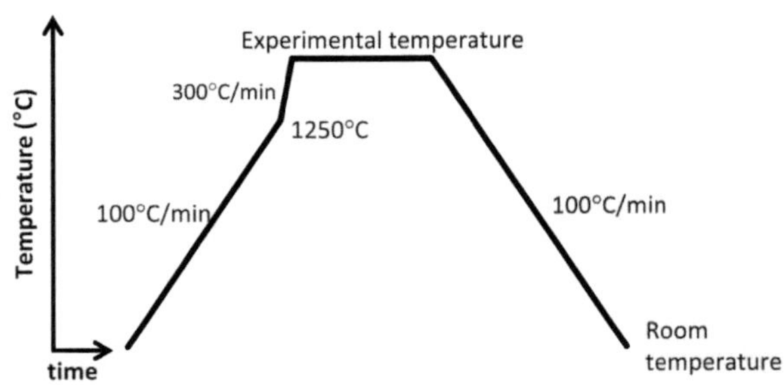

involving dissolution of 0.2 mm sapphire and 0.4 mm alumina–magnesia particles in slags 1 and 3 were quenched prior to complete dissolution, at 40–50% of total dissolution time. These slag–inclusion samples were cold-mounted, sectioned and polished to 1 μm for SEM-EDS analysis to investigate the reacting interface.

The heating profile used for dissolution experiments is shown in Fig. 2. It is noted that for the series of experiments which were arrested for electro-optical characterization, the cooling step is not as shown in Fig. 2, but the system was quenched from the experimental temperature to the room temperature.

The dissolution process in the LSCM was recorded to video, and the time and temperature were logged throughout the experiment. The video recordings from experiments were digitally processed and analyzed using image analysis software ImageJ to obtain inclusion dimensions with time. A border was drawn around the particle image which was then converted to an area measurement. A radius value was then calculated from the area considering the spherical shape of the particle. The radius of the particle was calculated for different reaction times. The use of high-precision spheres of inclusions lowers the errors associated with data analysis.

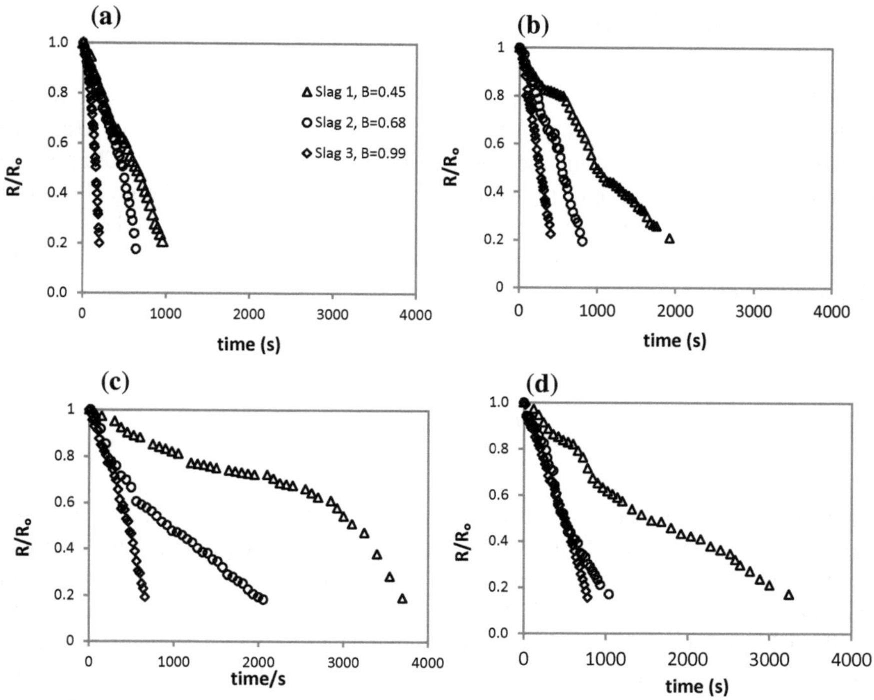

Fig. 3 Dissolution rate for sapphire particles of (**a**) 0.2 mm, (**b**) 0.3 mm and (**c**) 0.4 mm in slags 1–3 (**d**) dissolution rate for 0.4 mm alumina–magnesia particle in slags 1–3 (the particle size given is the nominal diameter of the spherical particles)

Results and Discussion

Dissolution Rate

The dissolution rate of sapphire and alumina–magnesia particles in slags 1–3 is given in Fig. 3. The dissolution rate of the particles is represented through relative change of the particle radius (R/R_0) with time, where R_0 is the particle radius at time zero and R is particle radius at time t. It can be seen that the rate of dissolution of both sapphire and alumina–magnesia particles increased when slag basicity increased from 0.45 for slag 1 to 0.99 for slag 3. Dissolution of 0.4 mm alumina–magnesia was faster than that of 0.4 mm sapphire in slags 1 and 2 but not in slag 3.

Dissolution Mechanism

The shrinkage core model (SCM) has been shown to be a good representation of inclusion dissolution in slags [12, 30–47]. In this study, SCM has been used to attempt to establish the rate-controlling mechanism of the systems studied. The SCM predicts the progressive dissolution of the particle in Stokes regime when chemical reaction and mass transfer in turn control the rate of dissolution. The normalized dissolution curves for chemical reaction and mass transfer models are obtained via Eqs. (1) and (2), respectively.

$$\frac{t}{\tau} = 1 - \frac{R}{R_0} \tag{1}$$

$$\frac{t}{\tau} = 1 - \left(\frac{R}{R_0}\right)^2 \tag{2}$$

where τ is the time for complete dissolution of the particle. The mechanism test then would be if the normalized dissolution curves fit with those predicted for chemical reaction control or mass transfer control. The predicted curves for chemical reaction and mass transfer models in the SCM (Eqs. 1 and 2) are compared to the experimental results in Fig. 4 for sapphire and the alumina–magnesia particle. The results of this test for sapphires of different initial sizes were not the same. The dissolution curves for 0.2 mm and 0.4 mm sapphires were generally in agreement with the mass transfer model, the exception being for the 0.4 mm particle in slag 2 (see Fig. 4a, c). In Fig. 4b, for the 0.3 mm particle though, the curves for slags 1–3 were in agreement with the chemical reaction model. The test conclusion for sapphire would be non-discriminating. Though classed as non-discriminating, the majority of the data were consistent with mass transfer in slag phase control. On the other hand, dissolution curves for the alumina–magnesia particle in Fig. 4d were generally in agreement with the chemical reaction model.

Slag–Inclusion Interface Assessment

The slag–particle interface SEM-EDS characterization with a view to aid establishing the reaction-controlling mechanism was carried out using JEOL JSM 6490LV system. EDS analysis was carried out using the JEOL EX-230xBu analyzer and an Oxford X-max^N 80 mm^2 SDD analyzer. The SEM-EDS analysis results of slag 1-inclusion interface are given in Figs. 5 and 6 for sapphire and the alumina–magnesia particle, respectively. The similar analysis was conducted for slag 3 and inclusion interface. The particle compositions in the central region are consistent with the original particle composition. EDS line analysis showed that there is a concentration gradient of elements (particularly Al and Si) between the slag–particle interface and the bulk slag for both sapphire and alumina–magnesia particles in slags tested.

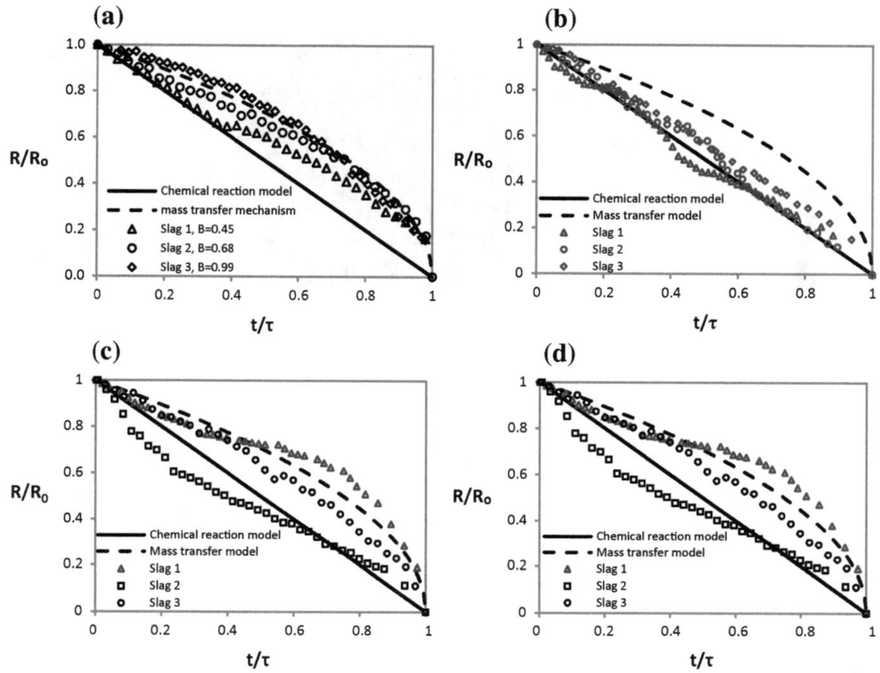

Fig. 4 Progress of reaction for (**a**) 0.2 mm sapphire, (**b**) 0.3 mm sapphire and (**c**) 0.4 mm sapphire particles in slags 1–3 at 1500 °C in terms of time for complete reaction (d) Progress of reaction for 0.4 mm alumina–magnesia particles in slags 1–3 at 1500 °C in terms of time for complete reaction; results are compared to the kinetic models

For the sapphire particle–slag systems, Fig. 5, there appears to be an enriched Al layer around the particle–slag interface on the slag side of the interface. Is this evidence of a "stagnant" layer as expected in a mass transfer-controlled reaction? Perhaps, though it is difficult to be certain as the layer may also be a result of precipitation of an alumina-rich phase on cooling.

For the alumina–magnesia particle–slag system, the appearance of the interface for slag 1 (Fig. 6) is distinctly different from that of the sapphire particle and slag. Just below the surface of the alumina–magnesia particle in the slag 1, a two-phase layer was formed with an enriched Mg phase and a pure alumina phase (see Fig. 6). In this system, it was not clear whether a concentration gradient exists in the slag phase. The higher Mg level at the particle edge is a result of the line scan passing through the Mg-rich area of the solid particle.

Is this indicative of a reaction mechanism? The lack of concentration gradients in the slag phase is consistent with mass transfer in the slag phase not being controlling. Does it indicate chemical reaction control? Perhaps though there is the possibility of a diffusion-controlled reaction associated with Mg or Al transport within the particle. If it was the case, then there would be an expectation of concentration gradients

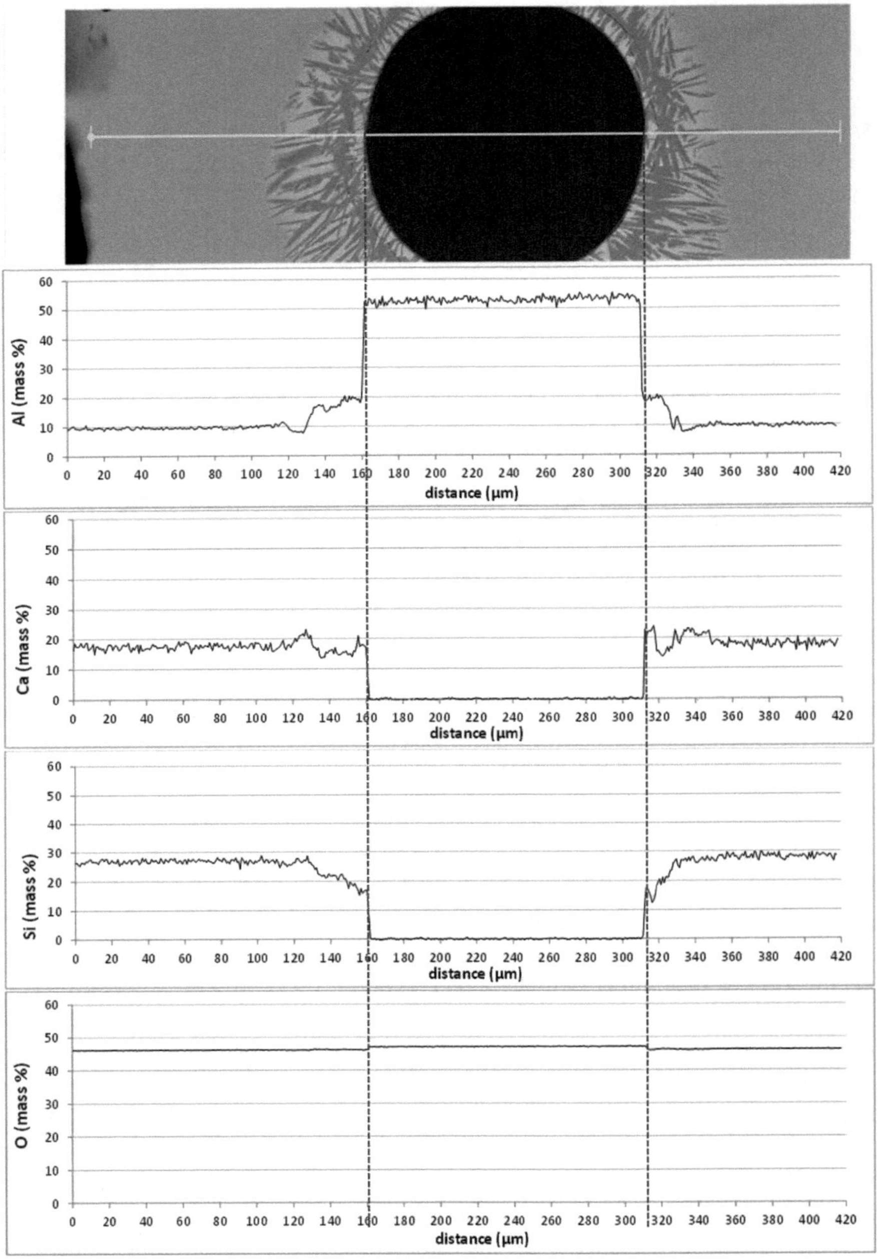

Fig. 5 EDS line analysis of the sapphire–slag 1 interface

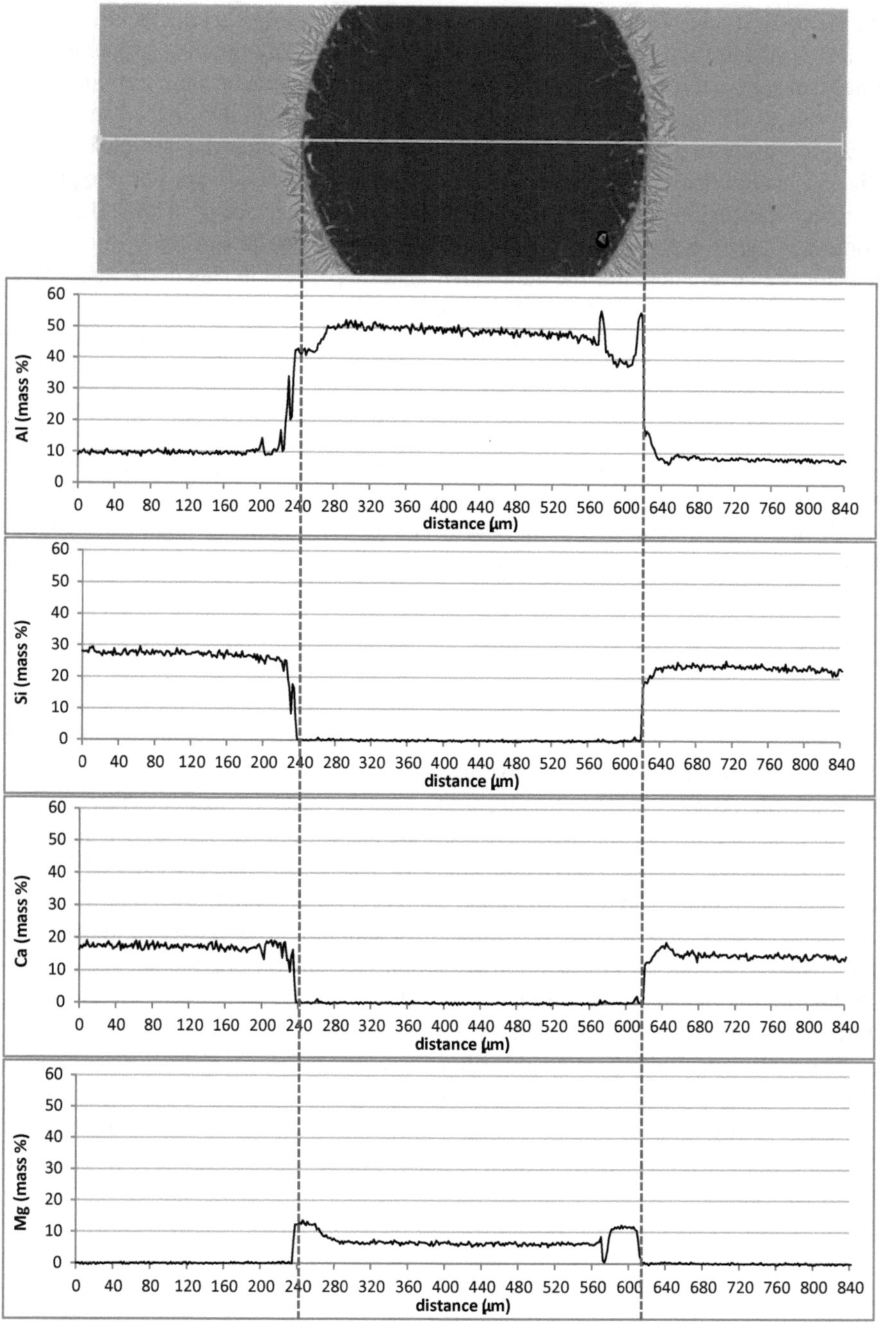

Fig. 6 EDS line analysis of the alumina–magnesia particle–slag 1 interface

within the particle. The spatial resolution of the Mg and Al curves is similar to that of any concentration gradient shown in the solid side. Any possible gradient in the line scans appeared steep (Fig. 6). Passing through the acicular pure alumina phase gave rise to disruptions in the elemental line scans, which also helped to obscure any concentration gradients. Therefore, it is not clear whether the particle shown in Fig. 6 is a chemical reaction controlled at the slag–particle or intra-particle product interface. Diffusion control in the particle side of the interface or chemical reaction control in the two-phase layer (Mg-rich and Al-rich) on the surface of the particle has not been evaluated. For the alumina–magnesia particle in slag 3, the appearance is similar to that of the sapphire particle and therefore consistent with mass transport in the slag phase. It may be for the alumina–magnesia particle system the reaction mechanism changed in changing from slag 1 to 3.

In summary, though not definitive, the physical characterization of the particle–slag interface for the sapphire particle–slag system was broadly consistent with mass transfer control in the slag phase reaction system. Under the assumption of the aluminum–oxygen complex anion being the rate-limiting species of the particles, the diffusion coefficient of aluminum–oxygen complex anion through the boundary layer can be obtained from SCM (Eq. 3):

$$D = \frac{R_0^2 \rho_B \left(\frac{B}{t}\right)}{2b(-\Delta C)} \tag{3}$$

where the $\left(\frac{R}{R_0}\right)^2 - 1$ term is defined as B, b is the stoichiometric coefficient of particle B, ΔC term is the driving force for reaction in molar units and equals difference between the Al_2O_3 interfacial concentration at the particle–slag interface and the Al_2O_3 concentration in slag bulk. The saturated slag compositions were established using MTDATA [49] thermodynamic analysis. The density and viscosity of the bulk and saturated slags 1–3 were obtained via NPL Slag Model [48]. The D values calculated via Eq. (3) are given in Table 2 for sapphire. The values varied between 10^{-11} and 10^{-10} m^2/s. For same size particles, diffusion coefficient values were slightly higher in slag 1 compared to slag 3. The obtained values were in reasonable agreement with other reported data.

The authors would recommend further characterization for diffusion or chemical reaction within the particle as an obvious area that requires further delineation and should be followed up in future studies of these slag–inclusion systems; in particular, dissolution experiments at different temperatures and different convection condition of the slag phase.

Table 2 Calculated D for sapphire from the SCM using Eq. (3)

Slag	Nominal diameter (mm)	R_0 (μm)	D (m^2/s)
1	0.2	93.0	1.29×10^{-10}
	0.3	135.0	1.36×10^{-10}
	0.4	193.1	9.24×10^{-11}
2	0.2	95.8	6.61×10^{-11}
	0.3	138.3	1.12×10^{-10}
	0.4	205.4	1.14×10^{-10}
3	0.2	91.6	9.66×10^{-11}
	0.3	145.2	1.42×10^{-10}
	0.4	191.2	1.46×10^{-10}

Conclusions

The focus of this study was an attempt to establish dissolution kinetics of sapphire and alumina–magnesia (similar to spinel) particle using high-precision spheres of the inclusions in LSCM technique. The rate data showed that dissolution of both sapphire and alumina–magnesia particles increased when slag basicity increased from 0.45 for slag 1 to 0.99 for slag 3. From the kinetic analyses carried out, a definitive conclusion on the rate-controlling mechanism of the sapphire into the $CaO–Al_2O_3–SiO_2$ slags tested cannot be made. Notwithstanding this point, there are indications that it may in part be mass transfer in slag phase controlled. Physical evidence in support of mass transfer mechanism for sapphire was an Al-enriched stagnant layer characterized around the dissolving sapphire particle in slags 1 and 3. Review of the kinetic analyses of the rate data for the alumina–magnesia particles would indicate on balance that the rate of dissolution of the alumina–magnesia particle is in part limited by the chemical reaction at the particle–slag interface. When characterizing the reaction interface for the alumina–magnesia particle, there was no such concentration gradient around the particle in slag 1, supportive of reaction mechanism. In slag 3 though, there were characteristics similar to that of the sapphire particle and consistent with mass transfer in slag phase. The measured diffusion coefficient for sapphire in slags tested in this study ranged from 10^{-11} to 10^{-10} $m^2\ s^{-1}$ at 1500 °C.

Acknowledgements The authors acknowledge the support of BlueScope and the use of the Australian Research Council funded JEOL-JSM 6490 LV SEM at the UOW Electron Microscopy Centre.

References

1. Lowe JH, Mitchell A (1995) Clean steel. Institute of Materials, London, p 223
2. Deo B, Boom R (1993) Fundamentals of steelmaking metallurgy. Prentice Hall International, New York, p 254
3. Seetharaman, S (Ed) (2008) Fundamentals of metallurgy. Woodhead Publishing in Materials Cambridge, England, p 23
4. Linder S (1974) Scand J Metall 3:137–150
5. Miki Y, Thomas BG (1999) Metall Mater Trans B 30B:639–654
6. Hallberg M, Jönsson PG, Jonsson TLI, Eriksson R (2005) Scad J Metall 34:41–56
7. Strandh J, Nakajima K, Eriksson R, Jönsson P (2005) ISIJ Int 45:1597–1606
8. Shannon G, White L, Sridhar S (2008) Mater Sci Eng, A 495:310–315
9. Strandh J, Nakajima K, Eriksson R, Jönsson P (2005) ISIJ Int 45:1838–1847
10. Sichen D (2012) Steel Res Int 83:825–841
11. Seetharaman, S (ed) (2008) Fundamentals of metallurgy. Woodhead Publishing in Materials Cambridge, England, pp 23–30
12. Valdez M, Prapakorn K, Cramb AW, Sridhar S (2001) Ironmak Steelmak 29(1):47–52
13. Yu X, Pomfret RJ, Coley KS (1997) Metall Mater Trans B 28(2):275–279
14. Taira S, Nakashima K, Mori K (1993) ISIJ Int 33(1):116–123
15. Goto K, Argent BB, Lee WE (1997) J Am Ceram Soc 80:461–471
16. Sarpoolaky H, Ghanbari-Ahari K, Molin J, Lee WE (2001) Stahl Eisen 122:130–133
17. Sandhage KH, Yurek GJ (1990) J Am Ceram Soc 73:3633–3642
18. Sandhage KH, Yurek GJ (1990) J Am Ceram Soc 73:3643–3649
19. Sandhage KH, Yurek GJ (1988) J Am Ceram Soc 71:478–489
20. Sandhage KH, Yurek GJ (1991) J Am Ceram Soc 74:1941–1954
21. Ueda K (1999) Mater Trans, JIM 63:989–993
22. Lee MS, Wright S, Jahanshahi S (2001) Metall Mater Trans B 32B:25–29
23. Dick AF, Yu X, Pomfret RJ, Coley KS (1997) ISIJ Int 37(2):102–108
24. Chung YD, Schlesinger ME (1994) J Am Ceram Soc 77:611–616
25. Xiang S, Lv X, Yu B, Xu J, Yin J (2014) Metall Mater Trans B 45B:2106–2117
26. Amini SH, Brungs MP, Jahanshahi S, Ostrovski O (2006) Metall Mater Trans B 37B:773–780
27. Taira S, Machida A, Nakashima K, Mori K (1996) Tetsu-to-Haganè 82:99–104
28. Choi JY, Lee HG, Kim JS (2002) ISIJ Int 42(8):852–860
29. Cho, WD, Fan, P (2002) EPD cong. and fundamentals of advanced materials for energy conversion, Seatlle, WA, USA, TMS, pp 631–638
30. Sridhar S, Cramb AW (2000) Metall Mater Trans B 32(2):406–410
31. Monaghan BJ, Chen L (2004) J Non-Cryst Solids 347:254–261
32. Monaghan BJ, Chen L (2005) Steel Res Int 76:348–354
33. Monaghan BJ, Chen L, Sorbe J (2005) Ironmak Steelmak 32(3):258–264
34. Monaghan, BJ, Nightingale, SA, Chen, L, Brooks, GA (2004) Molten 2004 conference, Johannesburg, South Africa, SAIMM, pp 585–594
35. Lee SH, Tse C, Yi KW, Misra P, Chevrier V, Orrling C (2001) J Non-Cryst Solids 282:41–48
36. Fox AB, Valdez ME, Gisby J, Atwood RC, Lee PD, Sridhar S (2004) ISIJ Int 44(5):836–845
37. Soll-Morris H, Sawyer C, Zhang ZT, Shannon GN, Nakano J, Sridhar S (2009) Fuel 88:670–682
38. Liu J, Verhaeghe F, Guo M, Blanpain B, Wollants P (2007) J Am Ceram Soc 90(12):3818–3824
39. Michelic S, Goriupp J, Feichtinger S, Kang YB, Bernhard C, Schenk J (2016) Steel Res Int 87(1):57–67
40. Valdez M, Prapakorn K, Cramb AW, Sridhar S (2001) Steel Res 72(8):291–297
41. Tse C, Lee SH, Sridhar S, Cramb AW (2000) 83rd steelmaking conference, Pittsburgh, PA, USA, pp 219–229
42. Monaghan BJ, Chen L (2006) Ironmak Steelmak 33(4):323–330
43. Yi KW, Tse C, Park JH, Valdez M, Cramb AW, Sridhar S (2003) Scand J Metall 32:177–184
44. Feichtinger S, Michelic SK, Kang YB, Bernhard C (2014) J Am Ceram Soc 97:316–327

45. Park JH, Park JG, Min DJ, Lee YE, Kang YB (2010) J Eur Ceram Soc 30:3181–3186
46. Levenspiel, O (1993) The chemical reactor omnibook. OSU Book Stores, Corvalis, OR, USA, pp 51.1–51.6
47. Levenspiel, O (ed) (1999) Chemical reaction engineering, 3rd edn. Wiley, New York, NY, USA, pp 566–588
48. Mills, KC (1991) Slags model (ed 1.07). National Physical Laboratory, UK
49. Davies RH, Dinsdale AT, Gisby JA, Hodson SM, Ball RGJ (1994) Applications of thermodynamics in the synthesis and processing of materials, Rosemont, IL, USA, pp 371–384

Part V
Additive Manufacturing and Biomaterials

In Situ Characterization of Hot Cracking Using Dynamic X-Ray Radiography

Po-Ju Chiang, Runbo Jiang, Ross Cunningham, Niranjan Parab, Cang Zhao, Kamel Fezzaa, Tao Sun and Anthony D. Rollett

Abstract We employ dynamic X-ray radiography (DXR) for in situ and real-time characterization of the hot cracking phenomenon for aluminum alloy 6061 under the processing conditions typical of laser powder bed fusion. The dynamics of processes such as a crack initiating from a bubble trapped subsurface are captured. We also directly observe the backfilling of liquid that heals an open crack. In addition, we demonstrate the feasibility of determining the point of origin for hot cracking with a temporal resolution of order 20 μs and spatial resolution of order 2 μm. This could shed light on the estimation of solid fraction at the initiation of hot cracking, which is a critical parameter upon which many models are based. We demonstrate the capability of DXR for generating new insights into verify or refine hot cracking models, and understand this problem fundamentally, which could ultimately lead to the optimization of process control for additive manufacturing.

Keywords Hot cracking · Dynamic X-ray radiography · Additive manufacturing

Introduction

The phenomenon of cracking during solidification is known by multiple names such as hot cracking, hot tearing, and solidification cracking [1, 2]. The problem has long been identified in the fields of casting, welding, and additive manufacturing [3, 4]. Many models have been proposed to predict the susceptibility to hot cracking [1, 2, 5]

P.-J. Chiang · R. Jiang · R. Cunningham · A. D. Rollett (✉)
Department of Materials Science and Engineering, Carnegie Mellon University,
5000 Forbes Avenue, Pittsburgh, PA 15213-3890, USA
e-mail: rollett@andrew.cmu.edu

N. Parab · C. Zhao · K. Fezzaa · T. Sun
X-Ray Science Division, Advanced Photon Source,
Argonne National Laboratory, Argonne, IL 60439, USA

J. Nakano et al. (eds.), *Advanced Real Time Imaging II*, The Minerals, Metals & Materials Series, https://doi.org/10.1007/978-3-030-06143-2_8

or the conditions under which it occurs [1, 6]. Rappaz, Drezet, and Gremanud formulated the well-known RDG model by combining the uniaxial tensile deformation of coherent solid network and the shrinkage feeding of interdendritic liquid to the mushy zone, i.e., the semisolid region during solidification [6]. More recently, Kuo has proposed another criterion for solidification cracking that considers the separation of neighboring grains from each other [2]. Eskin et al. reviewed the extensive literature on hot cracking models with their varied considerations and proposed criteria [1]. Unlike the solidification process for some transparent organic materials, which can be observed in situ by optical microscope [7], in situ characterization of the solidification process for metals is much more challenging. Although the post-analysis of microstructure after solidification has generated some information regarding the hot cracking phenomenon, certain parameters upon which the above theories are based must inevitably be estimated or guessed at because of the lack of in situ data for crack nucleation [2, 6]. Real-time observation is also not achievable with post-analysis, which may limit our understanding of what truly happens during the solidification process. Recognizing the needs, Matsuda et al. designed a customized optical microscope setup incorporated into a tungsten inert gas arc welding apparatus to dynamically observe solidification cracking during the welding process [8]. Despite the new insights given by this approach, only surface observations can be made on the welds, whereas the initiation of cracks and other detailed events that occur beneath the surface could not be monitored. It is not until the recent developments of more powerful synchrotron sources, which vastly enhances the quality of X-ray imaging, does the in situ and real-time observation for solidification of alloys become more practical with sufficient temporal and spatial resolutions [9, 10]. The full-field imaging technique realized by synchrotron can be applied as high-speed absorption and phase contrast radiography (dynamic X-ray radiography or DXR) , computed tomography, and white-beam topography [9]. In DXR, disparate phases and density differences in the material are distinguishable mainly from their differences in absorption of X-rays. This can be employed for the study of the highly dynamic phenomena during solidification such as cracking and columnar dendritic growth [11]. Computed tomography employs computer algorithms to reconstruct a 3D image by rotation of the sample over 180° or 360° while continually acquiring transmitted images. Although the application is typically post-analysis of solidified structures [12], it could also be used in situ provided the rate of data acquisition is fast enough so that the change of microstructures during one scan is trivial [13, 14]. White-beam topography records the variation in intensity of an X-ray beam diffracted from the sample and is frequently used for the characterization of defects and distortions in alloys during solidification [15]. In this work, we employed DXR to demonstrate in situ characterization for the nucleation of hot cracking and real-time observation of backfilling for 6061 aluminum alloy (AA 6061) under typical processing conditions of laser powder bed fusion.

Experimental Section

The DXR experiments were performed at the 32-ID-B beamline of Advanced Photon Source at Argonne National Laboratory [16, 17]. The schematic of the setup is shown in Fig. 1. Polychromatic X-rays with the first harmonic energy of 24.4 keV were generated from the short-period (18 mm) undulator. A set of slits in front of the undulator controlled the size of the X-ray beam. The slow and fast shutters were used to reduce the time for X-ray exposure to avoid damage to the detection apparatus. The specimen was AA 6061 plate sample with dimensions about 50 mm long, 3 mm tall, and 1 mm thick. It was put inside a stainless steel chamber of argon atmosphere and kept at room temperature before laser scanning, with no preheating applied. X-rays pass through the thickness direction and illuminate an area that fully encloses the melt pool as well as the space above the metal. The laser beam, set at a power level that is a selected fraction of the maximum 520 W, penetrates the top edge of the specimen in a direction perpendicular to that of the X-rays, and can be scanned across the sample horizontally at a speed up to 1.2 m/s. In the present work, single pass of the laser at 0.7 m/s for a duration ~3.8 ms was employed. There are no specific measures applied to assist the cooling process after the laser is turned off. Temperature measurement in the region of interest is not feasible because of the small volume and high cooling rates, which are estimated to be of order 10^6 K/s. A high-speed camera was placed behind the sample chamber to record X-ray radiographs at a frame rate of 50 kHz. The exposure time for each image was 1 μs. All the DXR images were processed via custom designed "adapthisteq filter" in MATLAB™ to enhance the contrast so that the features of interest could be characterized more readily.

Results and Discussion

Figure 2a is a static DXR image taken when no more microstructural change observed after the laser passed. The hot cracking along with bubbles trapped in the solid is clearly evident as the brighter area in the image. The dynamics of crack nucleation and propagation were captured in situ and in real time by a sequence of DXR images. Figure 2b–f shows a sequence of DXR images that zoomed into the area of interest pointed out by the arrow in Fig. 2a in order to highlight a crack nucleation event. The selected area of interest was inside the melt pool from about frame 100 through frame 150, suggesting the dot pointed out in Fig. 2b as a bubble. The movement of the bubble was observable before it was pinned at the place as shown in Fig. 2c. After the melt pool edge, which is also assumed to be the solidification front, passed the area of interest, the crack was observed to nucleate and propagate from the bubble within 500 μs as shown in Fig. 2d through 2f. Note that only a fraction of cracks was observed to nucleate from bubbles.

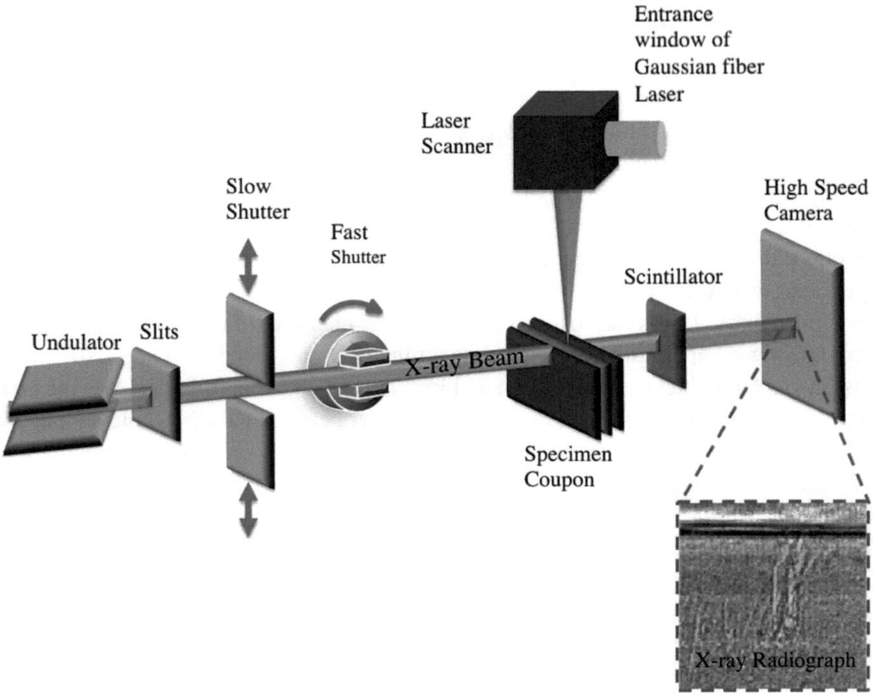

Fig. 1 Schematic of the dynamic X-ray radiography experiment [17] (reproduced with permission of the International Union of Crystallography)

For better quantification of the nucleation of hot cracking, an analysis procedure was employed and demonstrated as follows. First, an area was chosen as the crack of interest from the static X-ray radiograph as shown in Fig. 3a. The DXR images show that some cracks branched and entangled with each other. However, one could still identify an approximate area where the bulk of those cracks initiated by inspecting all frames of DXR images and choosing the frame that contains the first credible appearance of a crack. As depicted in yellow from Fig. 3b, the area inside the crack was determined by an algorithm written in MATLAB^{TM}. The algorithm aims to detect those points in the images whose intensities are above a selected threshold in the final state and have changed significantly compared to the first several frames. This is to ensure they were not artifacts initially present in the image such as the huge blurred white spot located in the rightmost of Fig. 2a (oxidized particles on the scintillator). The typical intensity threshold of cracks is chosen as roughly the value for bubbles with minor adjustment for each crack individually. For each point inside the crack, a plot for the change of its brightness was generated for inspection like the one shown in Fig. 3c. Some fluctuations in lower values were observed at the beginning before

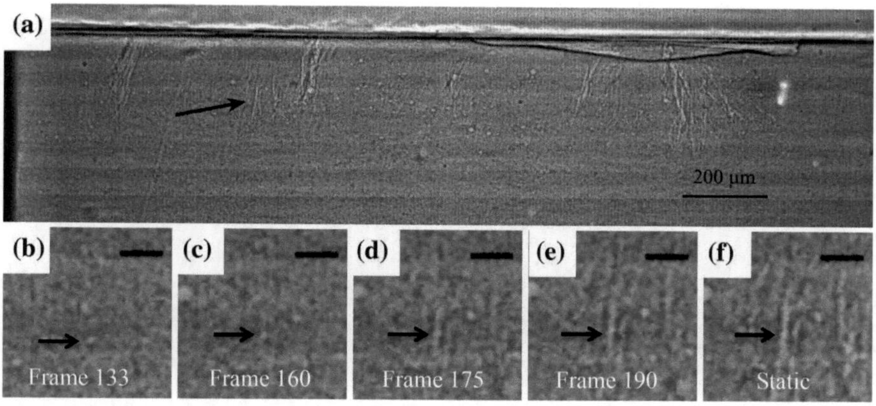

Fig. 2 A static X-ray radiographs showing the cracks after solidification and a sequence of zoom-in DXR images exhibiting real-time observation of cracks nucleating from a bubble. The scale bar is 30 μm for **b–f**. **a** The static X-ray radiograph taken after the sample fully solidified. **b** A bubble inside the melt pool was pointed out by the arrow. **c** The bubble was trapped in the solid after melt pool passed. **d** and **e** The nucleation and propagation of hot cracking from the bubble. **f** The static X-ray radiograph zoom into highlight the nucleated crack taken after the sample fully solidified

an abrupt increase delimited by the red and green lines. It is characteristic of the transformation from a higher-density phase of solid or liquid to a state of gas or cavity that absorbs less X-rays during the nucleation of cracks. The intensity of the crack point in the X-ray image remained high after the nucleation as no backfilling was observed in this case. It is worth mentioning that the passage of a bubble is also associated with an abrupt increase in intensity, but within a much shorter time frame, typically less than 40 μs. The end of the abrupt increase defined by the red line was chosen as the nucleation time for the material at the point. By comparing the nucleation time for all points inside the crack area, the point that transitioned the earliest is chosen as the nucleation point for the crack. Unsurprisingly, the points surrounding the nucleation point have similar nucleation times. Figure 3d displays the spatial characterization for the origin of crack nucleation.

The backfilling or back-diffusion phenomenon has been proposed by many to be the mechanism that could heal the cracks [18, 19]. However, no direct observation has been made to the authors' knowledge. Figure 4 shows a real-time observation of backfilling. A crack developed early after the passage of the melt pool, as detected by the increase in X-ray transmission. Subsequently, the area inside the crack shape darkened as the solidification progress advanced. The liquid phase does not exhibit a significant difference in brightness compared with the solid phase in DXR images. Therefore, we believe this suggests a decrease in material density as a crack opens up, followed by an increase as liquid flows back in and heals the crack. Note that

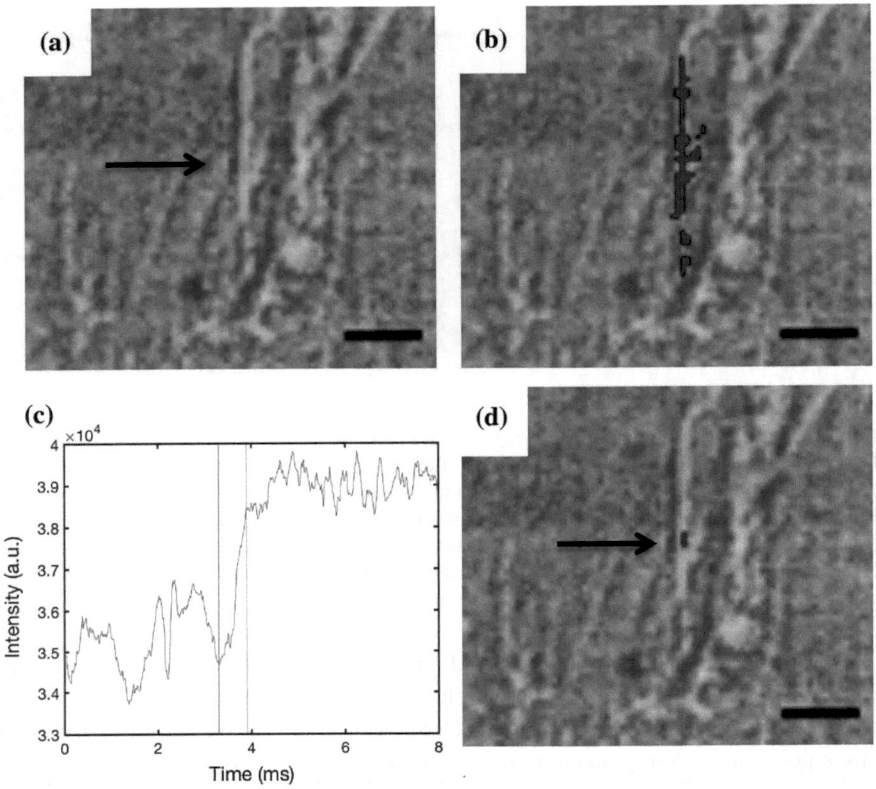

Fig. 3 Procedure for precise temporal and spatial characterization of crack nucleation. The scale bar is 30 μm in **a**, **b**, and **d**. **a** The crack of interest was chosen. **b** The area inside the crack was detected and labeled in yellow with the help of an algorithm programmed in MATLAB™. **c** The change of intensity for a point inside the crack during DXR experiment plotted versus time. The time depicted by the red line was the time for the material at that point to transform from a higher-density phase of solid or liquid to a state of gas that absorbs less X-rays. **d** The point of origin for hot cracking was determined and labeled in yellow

the backfilling liquid flowed from the top to the bottom of the crack, indicating the flow direction was from the solidification front toward the roots of primary dendrites, which is compatible with most models [2, 6]. The darker shade for the backfilled crack was probably because it solidified with the last fraction of liquid, which is expected to have higher solute content from segregation during the solidification of the alloy.

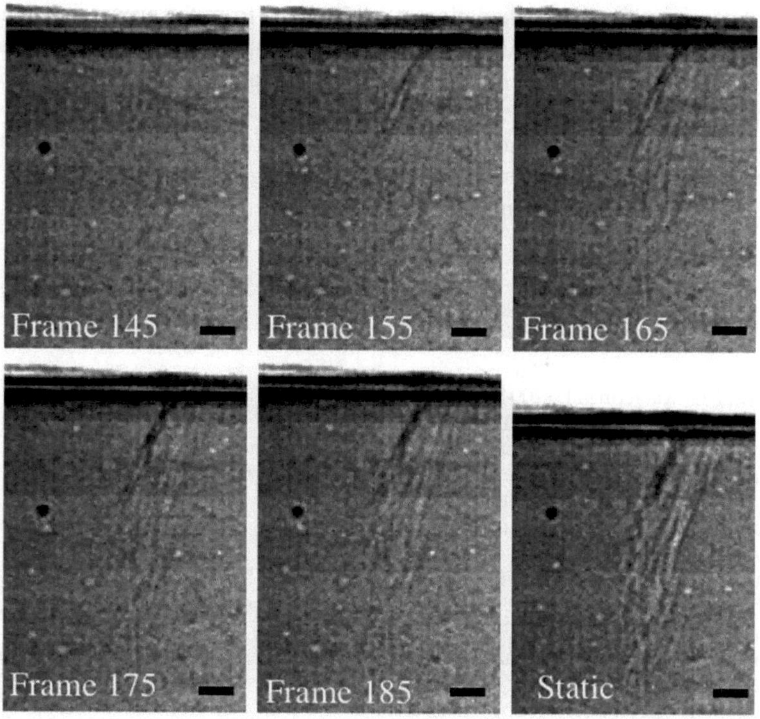

Fig. 4 A sequence of DXR images demonstrating the in situ observation of backfilling, based on a decrease in material density as a crack opens up, followed by an increase as liquid flows back into the crack. The scale bar is 30 μm

Conclusion

In summary, our work has demonstrated the use of DXR for in situ and real-time characterization of hot cracking under the processing conditions typical of laser powder bed fusion. We believe this capability will help generate new insights concerning this long-standing problem. The process for a crack initiating from a bubble trapped subsurface along with the backfilling phenomenon was directly observed. A procedure was designed to demonstrate the feasibility of determining the point of origin for hot cracking in excellent temporal and spatial resolutions. The precision is afforded by the frame rate of camera recording the DXR images, i.e., 50 kHz for our present work and higher speeds in future study. This enables the estimation of the solid fraction at crack nucleation point with the help of the correlation between temperature and solid fraction. The temperature at the point of crack initiation could be approximated by either calculation of cooling rate from the processing condition or simulation of the temperature profile. Since many models regarding hot cracking are highly dependent on the hypothesized or speculated values for the solid fraction upon crack nucleation [1, 2, 6], the direct characterization of crack nucleation points will contribute to refinement or verification of those models. Based on that, new insights may be discovered for the hot cracking phenomenon, leading to the development and optimization of process control for additive manufacturing.

References

1. Eskin DG, Katgerman L (2004) Mechanical properties in the semi-solid state and hot tearing of aluminium alloys. Prog Mater Sci 49(5): 629–711
2. Kou S (2015) A criterion for cracking during solidification. Acta Mater 88:366–374
3. Martin JH, Yahata BD, Hundley JM, Mayer JA, Schaedler TA, Pollock TM (2017) 3D printing of high-strength aluminium alloys. Nature 549(7672):365
4. Kou S (2003) Solidification and liquation cracking issues in welding. J Miner Metals Mater Soc 55(6):37–42
5. Easton MA, Wang H, Grandfield J, Davidson CJ, Stjohn DH, Sweet LD, Couper MJ (2012) Observation and prediction of the hot tear susceptibility of ternary Al–Si–Mg alloys. Metall Mater Trans A 43A(9):3227–3238
6. Rappaz, M, Drezet, JM, Gremaud, M (1999) A new hot-tearing criterion. Metall Mater Trans A 30(2):449–455
7. Jackson KA, Hunt JD (1965) Transparent compounds that freeze like metals. J Metals 17(9):1031
8. Fukuhisa M, Hiroji N, Kazuhiko S (1982) Dynamic observation of solidification and solidification cracking during welding with optical microscope (I): solidification front and behavior of cracking (materials, metallurgy & weldability), transactions of JWRI
9. Nguyen-Thi H, Salvo L, Mathiesen RH, Arnberg L, Billia B, Suery M, Reinhart G (2012) On the interest of synchrotron X-ray imaging for the study of solidification in metallic alloys. Comptes Rendus Phys 13(3):237–245
10. Zhao C, Fezzaa K, Cunningham RW, Wen HD, De Carlo F, Chen LY, Rollett AD, Sun T (2017) Real-time monitoring of laser powder bed fusion process using high-speed X-ray imaging and diffraction. Sci Rep 7:11
11. Mathiesen RH, Arnberg L (2005) X-ray radiography observations of columnar dendritic growth and constitutional undercooling in an Al–30 wt%Cu alloy. Acta Mater 53(4):947–956
12. Cunningham R, Nicolas A, Madsen J, Fodran E, Anagnostou E, Sangid MD, Rollett AD (2017) Analyzing the effects of powder and post-processing on porosity and properties of electron beam melted Ti–6Al–4V. Mater Res Lett 5(7):516–525
13. Puncreobutr C, Phillion AB, Fife JL, Rockett P, Horsfield AP, Lee PD (2014) In situ quantification of the nucleation and growth of Fe-rich intermetallics during Al alloy solidification. Acta Mater 79:292–303
14. Terzi S, Taylor JA, Cho YH, Salvo L, Suery M, Boller E, Dahle AK (2010) In situ study of nucleation and growth of the irregular alpha-Al/beta-Al_5FeSi eutectic by 3-D synchrotron X-ray microtomography. Acta Mater 58(16):5370–5380
15. Reinhart G, Buffet A, Nguyen-Thi H, Billia B, Jung H, Mangelinck-Noel N, Bergeon N, Schenk T, Hartwig J, Baruchel J (2008) In-situ and real-time analysis of the formation of strains and microstructure defects during solidification of Al–3.5 wt pct Ni alloys. Metall Mater Trans A 39A(4): 865–874
16. Sun T, Fezzaa K (2016) HiSPoD: a program for high-speed polychromatic X-ray diffraction experiments and data analysis on polycrystalline samples. J Synchrotron Radiat 23:1046–1053

17. Parab ND, Zhao C, Cunningham R, Escano LI, Fezzaa K, Everhart W, Rollett AD, Chen L, Sun T (2018) Ultrafast X-ray imaging of laser–metal additive manufacturing processes. J Synchrotron Radiat 25(5):1467–1477
18. Lippold JC (2015) Welding metallurgy and weldability, welding metallurgy and weldability, pp 1–400
19. Liu JW, Duarte HP, Kou S (2017) Evidence of back diffusion reducing cracking during solidification. Acta Mater 122:47–59

High-Frequency Ultrasound Analysis in Both Experimental and Computation Level to Understand the Microstructural Change in Soft Tissues

Leila Ladani, Koushik Paul and Jeremy Stromer

Abstract High-frequency ultrasound has become a popular tool in characterizing small-scale soft materials. This method is particularly effective in pitch-catch mode. It has been used in tissue phantoms to evaluate the microstructure. This method has the potential to be used in determining the tissue pathology in cancer and other tissue degenerative diseases. Among different types of parameters of ultrasound, peak density has been found to be most sensitive to the microstructural and scatterer variations in soft materials. 25 MHz ultrasound wave is used in a pitch-catch mode to evaluate mm scale tissue phantoms with microscale scatterers on different thickness levels. FFT is used to convert the receiving signal to frequency domain, calibrate to remove the noise and analyze the peak density. Finite element simulation is used to model the wave propagation in the medium containing scatterers to find a systematic correlation with the scatterer density.

Keywords High-frequency ultrasound · Pitch-catch · Peak density
Gelatin phantom · Finite element analysis

Introduction

Ultrasound analysis has become one of the most familiar nondestructive testing methods around the globe. It is a very useful tool to understand the internal structural change of material as well as to visualize it. One of the key advantages of using ultrasound is its higher penetration depth compared to other nondestructive testings.

L. Ladani (✉) · K. Paul
University of Texas at Arlington, Arlington, TX, USA
e-mail: leila.ladani@uta.edu

K. Paul
e-mail: koushik.paul@mavs.uta.edu

J. Stromer
University of Connecticut, Storrs, CT, USA
e-mail: jeremy.stromer@uconn.edu

J. Nakano et al. (eds.), *Advanced Real Time Imaging II*, The Minerals, Metals & Materials Series, https://doi.org/10.1007/978-3-030-06143-2_9

It is also safe and relatively inexpensive compared to the other noninvasive testing methods. For this reason, it has become very common in modern days medical applications [1, 2].

In the USA, breast cancer is the most common form of cancer among women. After lung cancer, it is the second leading cause of death for US women. In 2018, around 63,960 cases of in situ breast carcinoma and 266,120 new cases of invasive breast cancer among US women are expected to be diagnosed [3]. For this reason, breast cancer remains a significant cause of concern in public health. Over the period of time due to both emotional and esthetic benefit, breast conservation therapy (BCT) has become preferred treatment over modified radical mastectomy (MRM) although MRM offers better local control compared to BCT [4]. Breast-conserving therapy (BCT) is a combination of two different processes called lumpectomy and irradiation therapy. Lumpectomy is the surgery where the lump (breast tumor) along with some surrounding normal tissue (margin) is being removed from the breast. The margin is being sent to the pathologist for further evaluation. If abnormalities are found in the margin during the pathological test, the patient has to go through some additional surgery. As a result, to avoid local recurrence (LR), the goal is to obtain a negative (cancer-free) margin [5–8]. Apart from the available postoperative margin evaluation process, there are also some intraoperative processes which require the presence of a specialist pathologist [9].

Therefore, to understand the tissue/margin pathology through individual analytical measurement as well as to visualize the internal structure in a short period of time, ultrasound is a very good option.

Clinical ultrasounds are being used extensively to examine different organs such as the eye, liver, breast, and kidney in the human body [10]. It is shown in some studies that there is a strong dependence between the characteristics of the ultrasonic wave propagated through the tissues and histological features of tissues including cell structure, cell number density, tissue microstructure, and tissue heterogeneity [11, 12]. Therefore, it can be said that ultrasound has the potential to differentiate between normal and cancer-affected tissue [9].

Doyle et al. [13, 14] worked with different parameters of quantitative high-frequency ultrasound to evaluate tissue microstructure which are sound speed, attenuation measurement, spectrum, and cepstrum. It was shown that among all these parameters, peak density and slope of the Fourier transform of the ultrasonic spectra show a large variation in different types of pathologies. Later peak density and attenuation data were used to differentiate between benign and malignant tissue [15]. The statistical measurement was done in both per position and per specimen basis. In both cases, the result of the study showed potential for ultrasonic analysis as a rapid and inexpensive intraoperative BCS method.

In this study, the goal was to understand the change in number of peaks of the power spectrum in terms of microstructural change in the phantom. The microstructural change had been created by adding inclusion into the phantom. Also, since high-frequency ultrasound attenuates with increasing travel distance, the phantom thickness was also considered as a parameter for peak counting in this experiment. Ultrasound analysis was done on tissue mimicking phantom by using through transmission (pitch-catch) technique. First the signal was sent only through a small amount of acoustic gel and then the signal was sent through the phantom having gel on both sides. Later, the phantom signal was calibrated by dividing with the signal sent through the gel point by point. At the next step, power spectrum of the signal was obtained using fast Fourier transformation (FFT) and the number of peaks was counted using a MATLAB algorithm. This technique had been conducted on plain gelatin phantom and phantom with inclusion on different thickness levels. Later on, finite element analysis simulation was done to observe the change in number of peaks caused by the different numbers of assorted sized scatterers and compare it with the experimental result.

Experimental Setup

In this experiment, ultrasound analysis was being done on tissue mimicking phantom (gelatin phantom). The phantom was prepared from unflavored gelatine, sugar-free psyllium hydrophilic mucilloid powder, and water. The weight ratio for this mixture was 20:10:250 [16]. Since it was desired to have soft tissue-like scatter in the phantom, glass microbeads of 3–6 μm diameter and 2.5 gm/cc density were mixed with this phantom as inclusion. The mixing volumetric ratio for the microbeads was 0.01% (v/v). The ultrasound analysis was being conducted using two high-frequency (25 MHz) transducers (2 mm diameter, TransducerWorks, Centre Hall, PA, USA). The -6 dB bandwidth of the transducers is in the 22–41 MHz range. The experimental setups shown in Fig. 1 were comprised of a mixed domain oscilloscope (Tektronix, Inc. Beaverton, OR 97077, USA), pulse receiver system (UTEX Scientific Instruments Inc. Mississauga, Ontario L5L 1A3, Canada), and ultrasonic transducers.

Data Acquisition

The ultrasound signal was transmitted through the tissue mimicking phantom and acoustic gel by using a "through transmission" (pitch-catch) process. First, the signal was generated by the pulse receiver system. The parameters of the pulse while generating the signal are listed below in Table 1. The generated signal was sent through the phantom by one of the ultrasonic transducers. The signal is then received by another transducer and is sent to pulser for amplification. It was then sent to an oscilloscope for analysis. The steps of the experimental process are shown in Fig. 2.

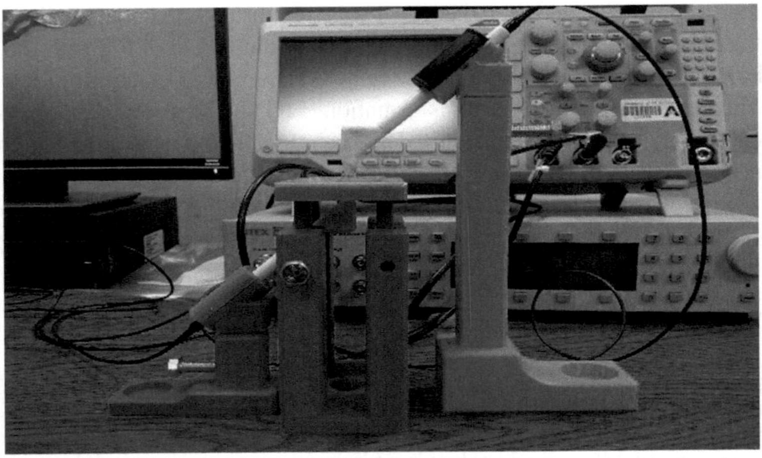

Fig. 1 Experimental setup for high-frequency ultrasound analysis

Table 1 Ultrasound signal parameter

Parameter	Value
Voltage	100 V
Pulse width	20 nS
Repetition rate	19 kHz
P/E gain	−6 dB
P/C gain	44 dB

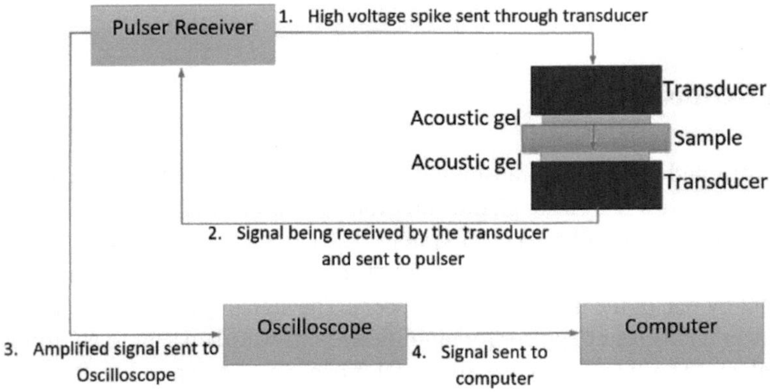

Fig. 2 Experimental process steps

Signal Processing and Peak Counting

From oscilloscope, the signal data was stored in the computer and later MATLAB program was used to post-process and analyze the signal. First, a rectangular window was used to isolate the pulse from the noise as in the signal shown in Figs. 3 and 4.

The rectangular function used in the signal processing is given below:

$$\mathrm{W}(n) = \begin{cases} 1, 420 \leq n \leq 690 \\ 0, n\langle 420, n\rangle 690 \end{cases} \quad (1)$$

where $\mathrm{W}(n)$ = time response and n = time step.

After windowing, the time domain signal was transformed into the frequency domain through fast Fourier transformation (FFT) function in MATLAB. The frequency response was taken from 0 to 80 MHz as shown in Fig. 5. Later, the signal has been calibrated by another FFT signal acquired from acoustic gel only through the above-mentioned steps. The calibrated signal is shown in Fig. 6.

Finally, a peak counting algorithm was used to count the peaks of the calibrated signal [17]. In the algorithm, threshold amplitude criteria were introduced to count the peaks which exceed a user-defined amplitude difference. The value of the threshold amplitude was set to 9.87×10^{-4}. This is done to avoid counting insignificant peaks that have a very low amplitude difference. These peaks can be caused by the noise in the signal.

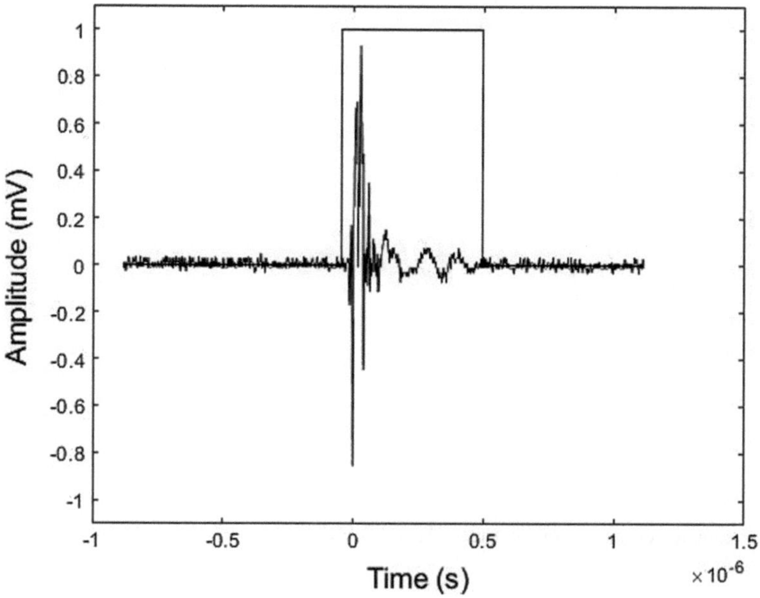

Fig. 3 Time domain signal with rectangular window

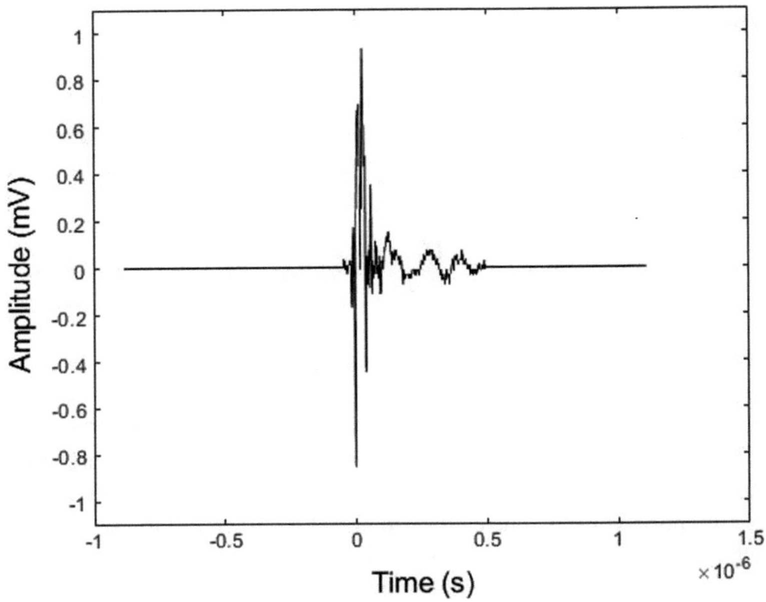

Fig. 4 Windowed time domain signal

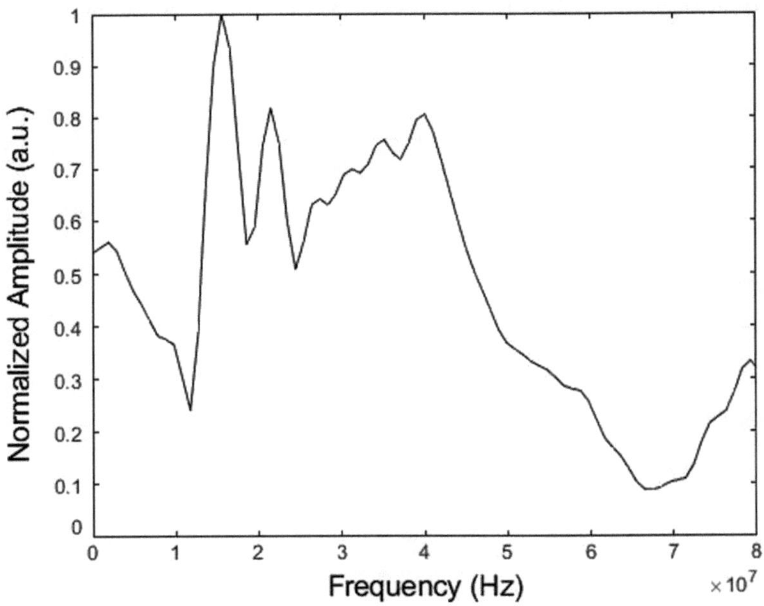

Fig. 5 Frequency domain signal between 0 and 80 MHz

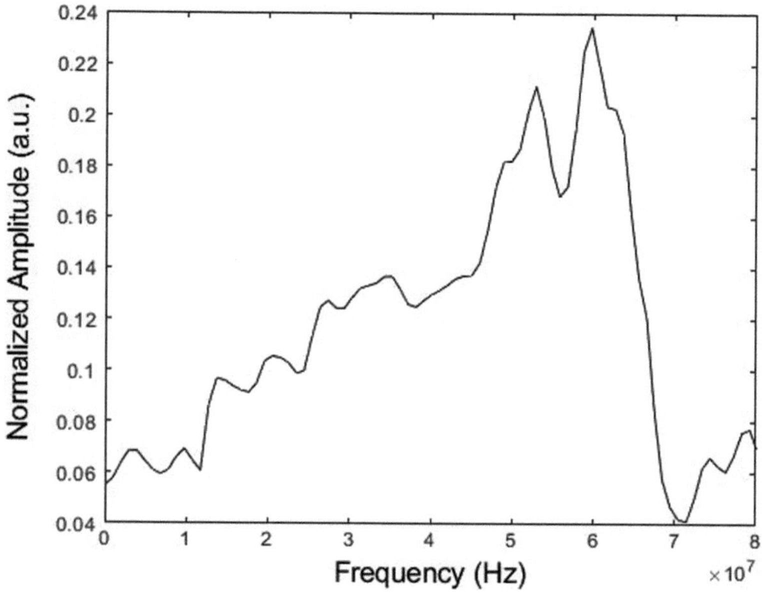

Fig. 6 Calibrated frequency domain signal between 0 and 80 MHz

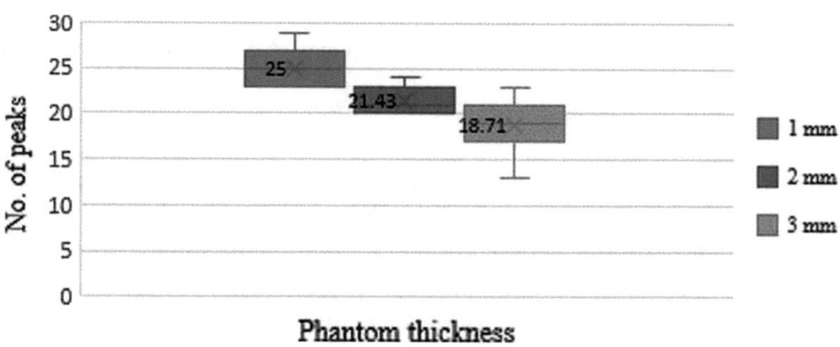

Fig. 7 Number of peaks for gelatin phantom with inclusion at different thickness level

Results and Discussion

In this study, peaks were counted in seven different positions of each phantom. For both the gelatin phantom with inclusion and plain gelatin phantom, the data were acquired from 1 to 3 mm thick samples and plotted in a box and whisker plot shown in Figs. 7 and 8.

From these above figures, it was evident that the number of peaks changes as the thickness of the phantom decreases. As a result, it could be concluded that the thickness of the phantom definitely played a role while the number of peaks changes. Also, in both cases, since the peak value had an increasing trend with decreasing

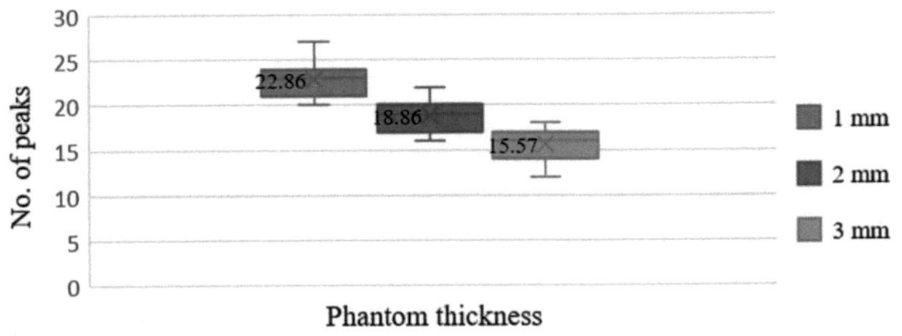

Fig. 8 Number of peaks for gelatin phantom without inclusion at different thickness level

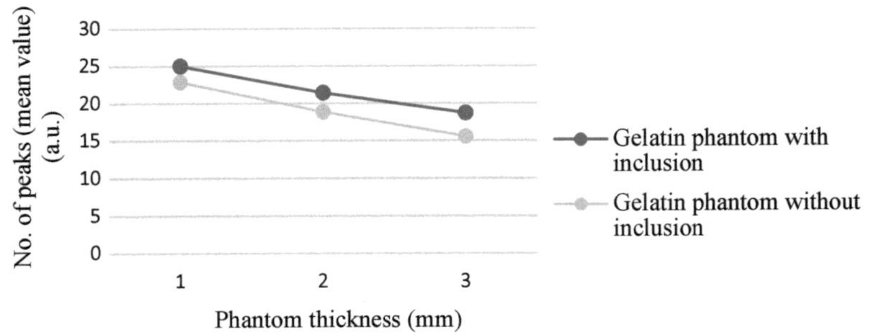

Fig. 9 Number of peaks versus phantom thickness for gelatin phantom with and without the inclusion

thickness, it could be stated that high-frequency ultrasound would give a higher number of peaks while using for a low thickness application. Later, to inspect the effect of added inclusion in gelatin phantom, mean peak values were taken from each sample of both types of phantom and plotted in Fig. 9. From this figure, it was observed that although the number of peaks increased while thickness decreased, the number of peaks on average was higher for samples with scatterers in all thicknesses.

From this observation, it could be stated that the tissue microstructure had an effect on the peak number along with the sample thickness. Therefore, in any high-frequency ultrasound application, if the sample thickness remains constant, it is reasonable to associate the number of peaks with the change in the scatterers density.

Finite Element Analysis (FEA) Simulation

The finite element model allows us to easily visualize the change in number of peaks caused by the different numbers of assorted sized scatterers. COMSOL Multiphysics v5.1 software was used to carry out the finite element simulations in this study. In the fluid region, COMSOL solves the 2D Helmholtz equation

$$\nabla \cdot \left(-\frac{1}{\rho_c} (\nabla p_t - \boldsymbol{q}_d) \right) - \frac{k_{eq}^2}{\rho_c} p_t = Q_m \tag{2}$$

where the equivalent wave number k_{eq} is defined for frequency and the fluid sound speed as c_c as

$$k_{eq}^2 = \frac{(2\pi f)^2}{c_c^2} - k_z^2 = k^2 - k_z^2 \tag{3}$$

The out-of-plane wave number k_z was set to zero in these models. Q_m and $\boldsymbol{q}_d$ correspond to monopole and dipole source terms. These terms were both zero in this system. The total pressure, p_t, is defined as the sum of the background pressure field and the scattered field: $p_t = p_b + p_s$. The background pressure field in these models is a plane wave propagating in the $\hat{e}_x$ direction and is given by $p_b = p_0 e^{-kx}$. Compiling these simplifications and using a constant fluid density, the equation became

$$\nabla^2 p_t + k^2 p_t = 0 \tag{4}$$

In order to fully capture the effects created by having elastic scatterers, the two-dimensional Navier's equation of motion was also solved for the solid scatterers. The microspheres were modeled as an isotropic, linear elastic material. The governing equation was then

$$-(2\pi f)^2 \rho_s \boldsymbol{u} = \nabla \cdot S + F e^{i\phi} \tag{5}$$

where f is the frequency, ρ_s is the density of the solid, S is the Cauchy stress tensor, F is the force acting on the solid, and ϕ is the phase component of the force. The displacement field of the solid is given by $\boldsymbol{u}$.

At the boundary of the fluid and the solid interface, the PDEs for the fluid acoustics and solid mechanics were coupled with the following boundary conditions

$$\boldsymbol{n} \cdot \frac{1}{\rho_c} \nabla p_t = -\boldsymbol{n} \cdot \boldsymbol{u}_{tt} \tag{6}$$

$$\boldsymbol{F}_A = p_t \boldsymbol{n} \tag{7}$$

where $\boldsymbol{n}$ denotes the unit vector normal to the boundary, $\boldsymbol{u}_{tt}$ is the structural acceleration, and $\boldsymbol{F}_A$ is the load acting on the solid structure.

The model was arranged into two simulation domains—fluid and solid. In addition, a perfectly matched layer (PML) surrounding the fluid domain was made. This layer mimicked a non-reflecting boundary condition and allowed any outgoing waves to leave the model domain without any reflection. A dense mesh was used to ensure proper resolution for the propagating waves, and the maximum element size was $\lambda/8$, where λ denotes the wavelength of the signal traveling through the medium.

In order to study the scattering due to different densities of scatterers, a layer with an area of 1 mm^2 and a height of 69 μm was used as well as the experimental

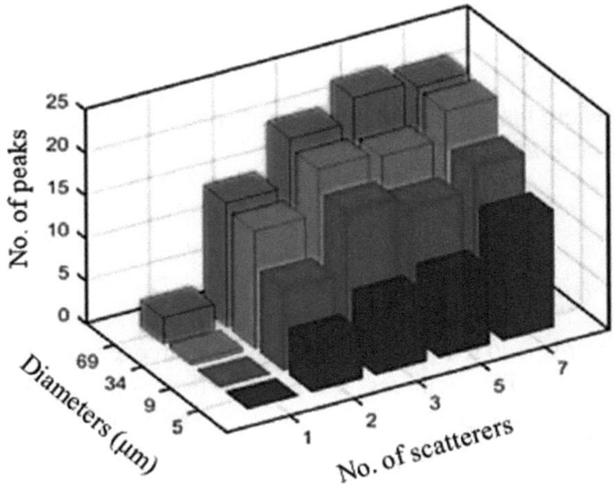

Fig. 10 Bar graph showing the measured number of peaks for the finite element model

number densities. The resulting number of scatterers used were 1, 2, 3, 5, and 7. Simulated peak densities were calculated for each scatterer geometry as well as for each scatterer diameter. The peak density values were averaged for each diameter and configuration over the four trials. To simulate the resulting experimental power spectra the simulation is repeated, and the Helmholtz equation is solved for each different k. In order to detect small changes in the resulting scattered pressure field for each frequency, the frequency was incremented in steps of 200 kHz and spanned the bandwidth of the transducer from 22 to 41 MHz. The scattered pressure field was determined on the back wall of the fluid modeling domain to obtain the overall forward response. After the simulation had been conducted for each wavenumber, the results were compiled into a model frequency spectrum. The derivative of the model spectrum was calculated, and the number of x-axis zero crossings were counted to calculate the simulated peak values. The peak values for the model are shown in Fig. 10.

Upon inspecting the graph, it is evident that the overall trend in which peak density tends to rise for increasing scatterer size as well as increasing numbers of scatterers. It appears that the peak density may be beginning to level off at large scatter size and high numbers of scatterers and there seem to be local variations in the overarching trend.

Conclusion

In this study, both experimental and computational work were focused on understanding the relationship between the high-frequency ultrasound wave peak density and the scatterers density in soft tissue phantoms. High-frequency ultrasound was sent through the tissues in a pitch-catch mode. The analysis was conducted on both plain gelatin phantom and phantom with inclusions. Additional experiments were conducted where phantom thickness effect on peak density was evaluated. It was observed that the peak value will increase if there are inclusions and scatterers exist in the phantom and if the thickness decreases. The effect of scatterers was observed as an independent factor from thickness. Also, finite element analysis showed that the peak value increases with added number of inclusions which matches the experimental result. In addition, computational analysis showed that the peak value will also increase if the inclusion size increases. Therefore, if the phantom thickness remains constant it is reasonable to state that the peak density will increase with increasing scatterer density.

References

1. Shull PJ (2002) Nondestructive evaluation: theory, techniques, and applications, New York, CRC Press, ISBN 9780824788728
2. Krautkrämer J, Krautkrämer H (1990) Ultrasonic testing of materials. Springer, Berlin, Heidelberg
3. U.S. Breast Cancer Statistics | Breastcancer.org (2018) https://www.breastcancer.org/symptoms/understand_bc/statistics. Accessed 11 Sep 2018
4. Litière S et al (2012) Breast conserving therapy versus mastectomy for stage I–II breast cancer: 20 year follow-up of the EORTC 10801 phase 3 randomised trial. Lancet Oncol 13(4):412–419
5. Pleijhuis RG, Graafland M, De Vries J, Bart J, De Jong JS, Van Dam GM (2009) Obtaining adequate surgical margins in breast-conserving therapy for patients with early-stage breast cancer: current modalities and future directions. Ann Surg Oncol 16(10):2717–2730
6. Jacobs L (2008) Positive margins: the challenge continues for breast surgeons. Ann Surg Oncol 15(5):1271–1272
7. Singletary SE (2002) Surgical margins in patients with early-stage breast cancer treated with breast conservation therapy. Am J Surg 184(5):383–393
8. Russo AL et al (2013) Margin status and the risk of local recurrence in patients with early-stage breast cancer treated with breast-conserving therapy. Breast Cancer Res Treat 140(2):353–361
9. Thill M, Baumann K, Barinoff J (2014) Intraoperative assessment of margins in breast conservative surgery—still in use? J Surg Oncol 110(1):15–20
10. Faverly DRG, Hendriks JHCL, Holland R (2001) Breast carcinomas of limited extent: frequency, radiologic-pathologic characteristics, and surgical margin requirements. Cancer 91(4):647–659
11. Mamou J, Oelze ML, O'Brien WD, Zachary JF (2005) Identifying ultrasonic scattering sites from three-dimensional impedance maps. J Acoust Soc Am 117(1):413–423
12. Wilke LG et al (2009) Rapid noninvasive optical imaging of tissue composition in breast tumor margins. Am J Surg 198(4):566–574
13. Doyle TE et al (2013) Determining breast pathology in surgical margins with high-frequency ultrasound: phantom and numerical simulations. J Acoust Soc Am 133(5):3541
14. Doyle TE et al (2011) High-frequency ultrasound for intraoperative margin assessments in breast conservation surgery: a feasibility study. BMC Cancer 11(1):444
15. Carter C, Neumayer LA, Factor RE, Doyle TE (2018) Abstract P6-03-10: using high-frequency ultrasound (20–80 MHz) to differentiate malignant vs benign breast tissue in surgical margins. Cancer Res 78(4 Suppl), P6-03-10 LP-P6-03-10
16. Bude RO, Adler RS (1995) An easily made, low-cost, tissue-like ultrasound phantom material. J Clin Ultrasound 23(May):271–273
17. Stromer J, Ladani L (2016) Investigating ultrasound imaging in the frequency domain for tissue characterisation. Nondestruct Test Eval 31(3):209–218

Part VI
Iron & Steelmaking III

Study of Mold Flux Heat Transfer Property by Using Thermal Imaging Enhanced Inferred Emitter Technique

Kaixuan Zhang, Wanlin Wang and Haihui Zhang

Abstract A thermal imaging enhanced inferred emitter technique was developed to investigate the heat transfer behavior of mold flux. Then, the phase transformation behavior, the heat transfer behavior, the temperature field evolution and the mold/slag interfacial thermal resistance evolution for a demonstration experiment of medium carbon mold flux slag disk were in situ recorded. The demonstration experiment results showed that the phase transformation behavior of mold flux significant affected the radiation heat transfer. And the phase transformation behavior also led to the change of temperature distribution on the slag. According to the in situ observation of slag temperature field, the crystallization behavior of mold flux made the high-temperature region move toward the crystalline layer. The variation of the mold/slag interfacial thermal resistance R_{int} also had been directly obtained with the help of thermal imager. R_{int} decreased with the increase of mold/slag interfacial temperature. In addition, mold/slag interfacial deformation and the decrease of interfacial temperature caused by the crystallization behavior led to an increase of R_{int}.

Keywords Mold flux · In situ · Heat transfer · Inferred emitter technique
Phase transition · Mold/slag interfacial thermal resistance

Introduction

Mold flux, one of the important materials for continuous casting, protecting the molten steel from oxidation, absorbing inclusions, lubricating the billet and controlling heat transfer in continuous casting production. Among them, the control of heat transfer directly affects to the quality of continuous casting billets. During the continuous casting process, mold flux is partially melted and covered on the surface of the molten steel in the mold, molten mold flux would infiltrate into the mold/shell

K. Zhang · W. Wang (✉) · H. Zhang
School of Metallurgy and Environment, Central South University,
Changsha 410083, Hunan, China
e-mail: wanlin.wang@gmail.com

J. Nakano et al. (eds.), *Advanced Real Time Imaging II*, The Minerals,
Metals & Materials Series, https://doi.org/10.1007/978-3-030-06143-2_10

gap and form slag film. The horizontal heat transfer from the liquid pool to the cooling mold mainly depends on the thermal physical properties of the slag film, such as crystallization properties, thermal conductivity, optical properties, mold/slag interface thermal resistance and so on [1–3]. Thus, the research on the thermal physical properties of mold flux is a key part in the process of continuous casting.

The major heat transfer mechanisms of slag film are heat conduction and thermal radiation. Susa et al. [4] and Hanao et al. [5], respectively, indicated the radiative heat flux constituted about 20% of the total heat for crystalline slag films, and Hanao also proved that the proportion reached 60% for purely glassy film. Some studies show that a part of the thermal radiation is scattered at the grain boundary [6], and the apparent reflectivity of slag film increases with crystallization behavior [5, 6], which implied that the crystalline layer has a more negative effect on radiation heat transfer than the purely glassy layer. However, Nishioka et al. [7] have proved that the crystalline slag film has better thermal conductivity than the glassy slag film. This means that the crystalline layer is more inclined to heat transfer by conduction. Besides, mold/slag interface thermal resistance (R_{int}) greatly influences the conduction heat transfer. And R_{int} increases with the degree of crystallinity of the slag film [3, 5, 8, 9].

Many experiment methods have been designed to study the thermal physical properties and heat transfer behavior of mold flux. For example, Takahashi et al. [10] studied the effective thermal conductivity by plane-parallel method. Wen et al. [11, 12] used a dip experiment to study the heat transfer and R_{int} evolution during the phase transition of mold flux. In experiments of Susa [4], Cho [13] and Xie [14], the optical properties of mold flux were investigated by the spectrophotometer. However, these techniques and methods are still unable to achieve in situ observation the phase transition process and the temperature field evolution process of slag films.

Inferred emitter technique [15, 16] (thermal imaging enhanced) was developed to investigate the heat transfer performance and the thermal properties of mold flux. This technique makes it possible to in situ observe the phase transformation process of slag films and in situ observation the temperature field evolution during the phase transformation process of mold flux film by using thermal imaging. Also, the mold/slag interface thermal resistance can also be obtained easily.

Experimental Apparatus and Procedure

A. Sample Preparation

The mold flux sample powder used in this study were prepared by prepared by mixing the powder of pure chemical reagents of CaO, SiO_2, Al_2O_3, MgO, CaF, Na_2O and Li_2O with size less than 0.074 mm (Supplier: Fine chemical engineering and technology research and development center, Guangdong, China). **Firstly**, the prepared sample powder was melted in an induction furnace at 1673 K for 300 s to homogenize the chemical compositions. **Secondly**, the melted mold flux was poured into a water-cooled copper mold to quench and achieve a fully glassy phase. The

Table 1 Compositions of the mold flux (wt pct)

CaO	SiO_2	Al_2O_3	MgO	F	Na_2O	Li_2O
43.06	34.44	4.00	2.00	8.00	7.50	1.00

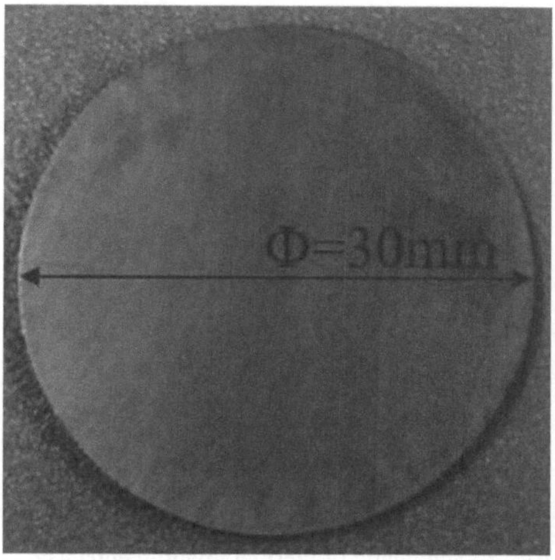

Fig. 1 Picture of the prepared fully glass

compositions of melted mold flux were analyzed by X-ray fluorescence (XRF) and listed in Table 1. **Thirdly**, the slag disk was annealed in a muffle furnace at the temperature of 873 K for 2 h to release the inner stress, and then cooled in the air. **Finally**, the slag disk was polished into a cylinder of 4.0 mm height and 30 mm diameter (Fig. 1) by using the SiC sandpapers with the grit size down to 1200.

B. Thermal Imaging Enhanced Inferred Emitter Technique Test

Figure 2 shows the schematic of thermal imaging enhanced inferred emitter technique (Ti-IET). Ti-IET consists of a water-cooled mold connecting with four T-type thermocouples, an infrared emitter, a refractor and an optical video camera [2]. Besides, Ti-IET is equipped with a thermal imager (Fluke Tix640 with 640 × 480 pixels) to measure the temperature field evolution of slag and mold during the experimental process. For a test ran, a slag disk was placed on the mold surface, and then the irradiation generated by the infrared emitter was applied on the slag disk surface so as to simulate the heat transfer of slag film in a continuous casting mold. While the profile of irradiative heat flux generated by infrared emitter in this study is shown in Fig. 3. During the test, the mold wall temperatures were measured by four thermocouples, crystalline phase transformation of the slag disk was recorded by an optical video camera, and the temperature of the slag disk was recorded by the thermal imager. To ensure the accuracy of temperature measurement by thermal imager, the emissivity (ε) was set as 0.82 that was obtained by comparing the slag temperature from a K-type thermocouple installed beneath the slag surface with the temperature from the thermal imager before the experiment.

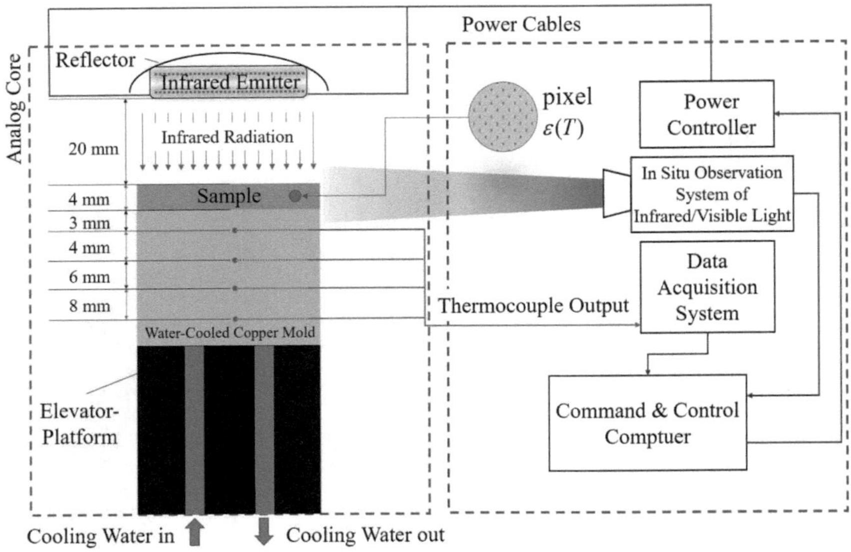

Fig. 2 Schematic of thermal imaging enhanced inferred emitter

Fig. 3 Incident radiative heat flux generated by infrared radiant emitter

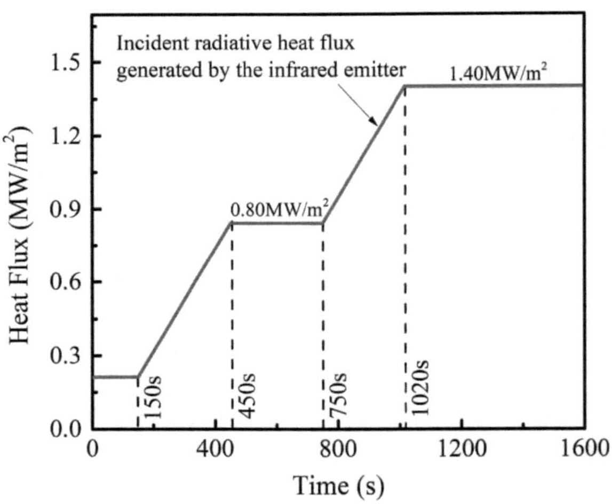

Results and Discussion

A. The Phase Transformation Behavior of Mold Flux

During the Ti-IET experiment, the heat flux (q_m) across mold top face was determined from the measured in mold temperatures by the famous Beck's 1D-IHCP [17]. Figure 4 shows the respond heat flux across the mold surface for the two test cases with/without a slag disk put on the mold top surface.

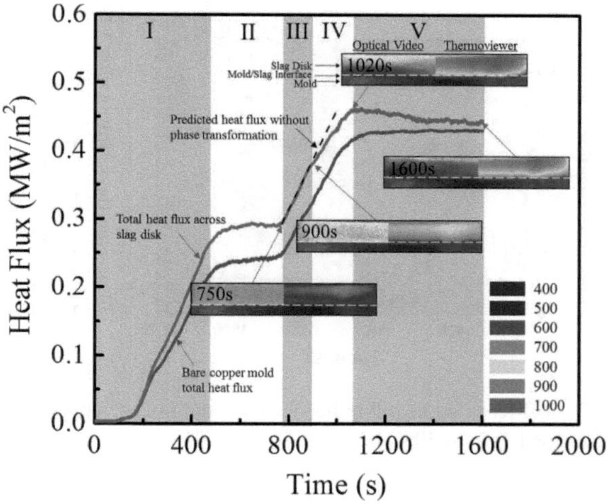

Fig. 4 Measured heat flux history and in situ recordings for slag disk

For the Ti-IET heat transfer test, the mold flux could be divided into five stages according to the total heat flux. In **Stage I** (0–450 s), both of mold heat fluxes for the bare mold test and the slag test increased from 0 to a relatively steady value with increasing incident radiative heat flux by IET lamp.

In **Stage II** (450–750 s), as incident radiative heat flux by IET lamp remained unchanged, the mold heat fluxes kept the relatively steady value of 0.24 and 0.29 MW/m^2 for the bare mold test and the slag test, respectively. According to the results of heat flux, we can find that the total heat flux across mold top surface of bare mold experiment was less than the case where the slag disk was placed. This phenomenon can be attributed to that the apparent reflectivity of mold top surface is bigger than that of glassy slag disk, when the glassy slag disk was placed on the mold top surface, the energy absorbed from the incident radiation was more than the bare mold. The slag was still pure glass in this stage.

In **Stage III** (750–900 s), both mold heat fluxes increased again due to the increase of incident radiation by IET lamp. At the beginning of **Stage IV**, the crystalline phase transformation of the slag disk was observed to occur at about 3.10 mm above the slag bottom. The slag crystallization will lead to the increase of the surface reflectivity and the scattering rate of slag disk, thereby, more radiation energy returned from the mold flux to the environment [18, 19]. This led to a decrease in the increase rate of mold heat flux although the increase rate of radiative heat flux by IET lamp remained unchanged in **Stage IV** (900–1020 s). This phenomenon is consistent with the previous study of heat transfer performance with different mold fluxes by using standard IET, as shown in Fig. 5a [18] and 5b [19], the previous results also suggested that the basicity tends to enhance the mold flux crystallization, leading to the reduction of radiative heat transfer rate and enlargement of interfacial thermal resistance. At the end of **Stage IV** (1020 s), the mold heat flux reached the maximum value of 0.42 and 0.46 MW/m^2 for the bare mold test and the slag test, respectively.

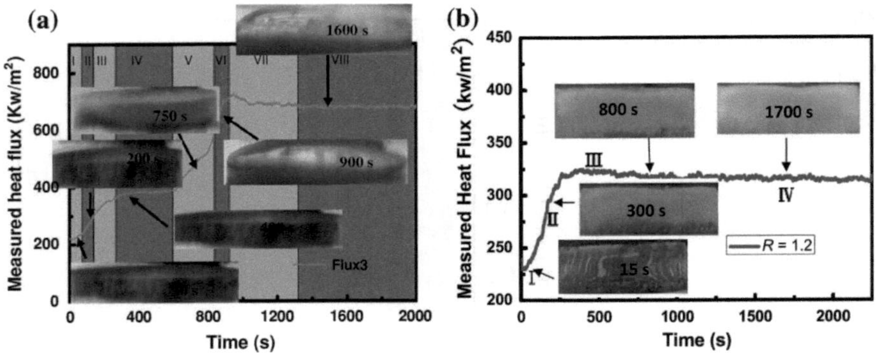

Fig. 5 Measured heat flux history for mold fluxes. **a** crystallization began at 735 s [18]. **b** crystallization began at 30 s [19]

In **Stage V** (1020–1600 s), the incident radiative heat flux reached the maximum of 1.40 MW/m^2 and remained for the rest of experiment. The mold heat flux for bare mold test also reached the maximum of 0.43 MW/m^2 and kept the relatively steady value for the rest of experiment. However, the mold heat flux for slag test reached its maximum value of 0.46 MW/m^2, then began to decrease with further slag crystallization; finally, the mold heat flux decreased to a relatively steady value of 0.44 MW/m^2 when the slag crystallization was completed. Moreover, according to in situ observation, the final range of crystalline layer is from 1.25 to 4 mm above the slag bottom.

B. The In Situ Observation of Temperature Field Evolution

During the Ti-IET experiment, the temperature field evolution of the slag disk was in situ recorded by the thermal imager, and temperature field diagrams at some key points are shown in Fig. 4. According to the temperature field data, the temperatures histories for different positions along the slag centerline in the vertical direction, and the slag disk surface temperatures by thermal imager from the slag bottom to slag top surface along the slag centerline in the vertical direction versus time are shown in Fig. 6a and b.

From Fig. 6a and b, it could observe that the position of the slag maximum temperature was not at the slag disk top surface. The position of slag maximum temperature was at 2.65, 2.66, 2.99, 3.13 mm above the slag bottom when the time was 750 s, 900 s, 1020 s and 1600 s, respectively. This phenomenon might be attributed to the convective heat transfer of the air and the thermal radiation from the slag to the ambient. According to the previous studies [6, 16], crystal has larger extinction coefficient than glass. When the phase transition occurred, the radiation scattered at grain boundaries would make a longer light path. Thus, a larger proportion incident radiative energy was absorbed or scattered by the crystalline layer, which located on the upper part of slag disk, and converted into its internal energy. This is the reason

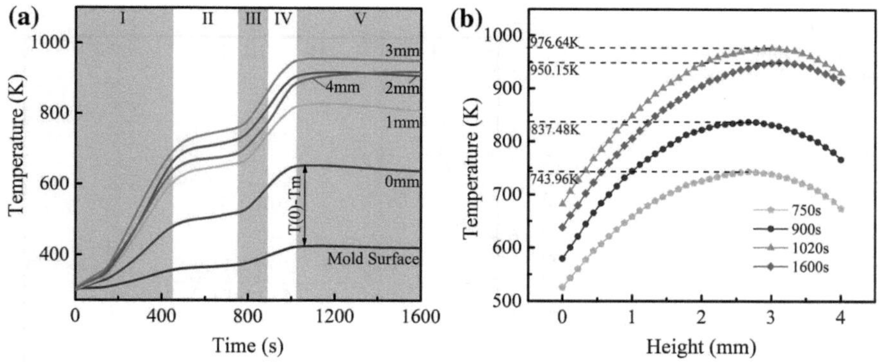

Fig. 6 **a** Temperatures for the different positions along the slag centerline in the vertical direction, and **b** Slag surface temperatures from the slag bottom to slag top surface along the slag centerline in the vertical direction at the different time

why the position of the slag maximum temperature rose with the crystallization behavior.

The mold/slag interfacial thermal resistance R_{int} could be calculated from the total heat flux q_m the temperature of slag disk bottom surface T_0 and the mold surface T_m.

$$T_m = T_1 + \frac{\lambda \cdot q_m}{z_0} \tag{1}$$

where $z_0 = 3$ mm is the distance from No. 1 thermocouple to the mold top surface, T_1 is an absolute temperature measured by No. 1 thermocouple, $\lambda = 384$ W/(m · K) is the copper thermal conductivity.

$$R_{\text{int}} = \frac{T_0 - T_m}{q_m} \tag{2}$$

where T_0 could be obtained from temperature field diagrams, T_m is calculated by Eq. (1). Combined with the total heat flux q_m (Fig. 4), the mold/slag interfacial thermal resistance R_{int} during the experiment could be calculated by Eq. (2). The calculation results of R_{int} are shown in Fig. 7.

According to the results of R_{int}, we can find that R_{int} was the largest of 6.88×10^{-4} m^2K/W at the start of the experiment, then, it reached the minimum of 5.44×10^{-4} m^2K/W when the slag surface was the hottest. This might be attributed to the increase of slag disk temperature and mold temperature lead slag disk bottom softening, and a closer contact between slag disk and mold. R_{int} increased slightly to 5.96×10^{-4} m^2K/W at 1600 s. This might be attributed to the increase of the interfacial thermal resistance between mold and slag introduced by the further crystallization of slag that might cause the deformation of the slag bottom surface and a worse contact between slag disk and mold.

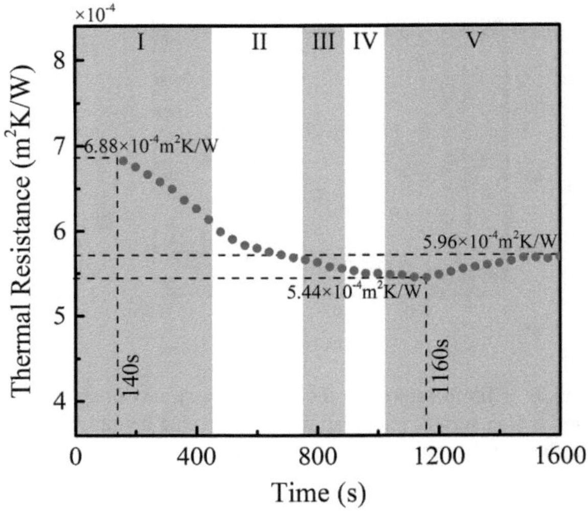

Fig. 7 Evolution of mold/slag interfacial thermal resistance during experiment

Conclusions

In this paper, the function and principle of thermal imaging enhanced inferred emitter technique (Ti-IET) have been emphatically demonstrated. The phase transformation behavior, heat transfer behavior, temperature field evolution and mold/slag interfacial thermal resistance evolution of medium carbon mold flux have been systematically studied by Ti-IET, and the main conclusions are summarized as follows:

(1) The phase transition behavior of mold flux has significant effect on the heat transfer behavior. The energy obtained from the incident radiation of the crystalline layer was less, because of the bigger apparent reflectivity, higher scattering rate, and the actual heat flux pass through the slag was less than the predicted heat flux without phase transition.
(2) The heat radiation from the mold flux to the environment and the convective heat transfer of the air made that the slag maximum temperature was not located at the top surface of slag disk, the position of slag maximum temperature moves up with the increase of crystallinity.
(3) The calculation results of mold/slag interfacial thermal resistance R_{int} shows that R_{int} was the largest of 6.88×10^{-4} m^2K/W at the start of the experiment, then, it decreased to the minimum of 5.44×10^{-4} m^2K/W with the increase of mold/slag interfacial temperature. A larger fraction of glassy slag became crystallinity that would lead to the increase of R_{int} by deformation of the disk.

Acknowledgements The financial supports from the National Natural Science Foundation of China (51704333, U1760202), and Newton Advanced Fellowship (NA150320) are great acknowledged.

References

1. Mills KC, Fox AB, Li Z, Thackray RP (2005) Performance and properties of mould fluxes. Ironmak Steelmak 32(1):26–34
2. Wang W, Cramb AW (2005) The observation of mold flux crystallization on radiative heat transfer. ISIJ Int 45(12):1864–1870
3. Yamauchi K, Sorimachi T Sakuraya, Fujii T (1993) Heat transfer between mold and strand through mold flux film in continuous casting of steel. ISIJ Int 33(1):140–147
4. Susa M, Kushimoto A, Endo R, Kobayashi Y (2011) Controllability of radiative heat flux across mould flux films by cuspidine grain size. ISIJ Int 51(10):1587–1596
5. Hanao M, Kawamoto M, Yamanaka A (2012) Influence of mold flux on initial solidification of hypo-peritectic steel in a continuous casting mold. ISIJ Int 52(7):1310–1319
6. Nakada H, Susa M, Seko Y, Hayashi M, Nagata K (2008) Mechanism of heat transfer reduction by crystallization of mold flux for continuous casting. ISIJ Int 48(4):446–453
7. Nishioka K, Maeda T, Shimizu M (2006) Application of square-wave pulse heat method to thermal properties measurement of $CaO–SiO_2–Al_2O_3$ system fluxes. ISIJ Int 46(3):427–433
8. Shibata H, Kondo K, Suzuki M, Emi T (1996) Thermal resistance between solidifying steel shell and continuous casting mold with intervening flux film. ISIJ Int 36(Suppl):S179–S182
9. Cho JW, Emi T, Shibata H, Suzuki M (1998) Heat transfer across mold flux film in mold during initial solidification in continuous casting of steel. ISIJ Int 38(8):834–842
10. Takahashi S, Endo R, Watanabe T, Hayashi M, Susa M (2018) Mechanism of mild cooling by crystallisation of mould flux for continuous casting of steel—a view from apparent thermal conductivity under steep temperature gradient. ISIJ Int 58(5):905–914
11. Yang G, Wen Q Sun, Tang P (2017) Evolution of temperature and solid slag film during solidification of mold fluxes. Metall Mater Trans B 48(2):1292–1307
12. Yang G, Wen Q Sun, Tang P (2017) Investigation of the coupled conductive and radiative heat transfer of molten slag in a cylindrical enclosure based on the zonal method. Int J Heat Mass Tran. 110:523–538
13. Yoon DW, Cho JW, Kim SH (2017) Controlling radiative heat transfer across the mold flux layer by the scattering effect of the borosilicate mold flux system with metallic iron. Metall Mater Trans B 48(4):1951–1961
14. Diao J, Xie B, Wang N, He S, Li Y, Qi F (2007) Effect of transition metal oxides on radiative heat transfer through mold flux film in continuous casting of steel. ISIJ Int 47(9):1294–1299
15. Wang W, Zhou L, Gu K (2010) Effect of mold flux melting and crystal fraction dissolution on radiative heat transfer in continuous casting. Met Mater Int 16(6):913–920
16. Wang W, Cramb AW (2010) Study of the effects of the mold surface and solid mold flux crystallization on radiative heat transfer rates in continuous casting. Steel Res Int 81(6):446–452
17. Beck JV, Blackwell B, St. Clair CR Jr (1985) Inverse heat conduction: III-posed problems, 1st edn. Wiley, New York, pp 267–279
18. Gu K, Wang W, Wei J, Matsuura H, Tsukihashi F, Sohn I, Min DJ (2012) Heat-transfer phenomena across mold flux by using the inferred emitter technique. Metall Mater Trans B 43(6):1393–1404
19. Gu K, Wang W, Zhou L, Ma F, Huang D (2012) The effect of basicity on the radiative heat transfer and interfacial thermal resistance in continuous casting. Mater Trans B 43(4):937–945

Sub-rapid Solidification Study by Using Droplet Solidification Technique

Cheng Lu, Wanlin Wang and Chenyang Zhu

Abstract Droplet solidification technique is important with respect to the fundamental study of strip casting given the common conditions of direct contact between cooling mold and solidifying metal. In this study, an improved droplet solidification technique has been developed for the in situ observation of the sub-rapid solidification phenomena of metal droplets impinging onto the water-cooled copper substrate. The heat transfer rates were calculated by the inverse heat conduction program (IHCP), according to the responding temperatures' gradient inside the cooling mold. Meanwhile a charge coupled device (CCD) camera was placed beside the bell jar aimed to record the whole melting and solidification process of the steel sample, which also allowed the determination of the final wetting angel, during the dipping tests. Moreover, it was found that the heat transfer rate increased with decreasing final contact angle, which means better wetting condition between the liquid sample and the copper substrate.

Keywords Droplet solidification technique · Sub-rapid solidification
In situ observation

Introduction

Strip casting is the process of producing thin strip directly from liquid metal, in which the process of rolling, annealing, etc., has been removed. Compared with the traditional continuous casting process, strip casting has the advantages of low investment and low running cost [1, 2]. As the total solidification time in the strip casting is

C. Lu · W. Wang (✉) · C. Zhu
School of Metallurgy and Environment, Central South University,
Changsha 410083, China
e-mail: wanlin.wang@gmail.com

C. Lu · W. Wang · C. Zhu
National Center for International Research of Clean Metallurgy,
Central South University, Changsha 410083, China

J. Nakano et al. (eds.), *Advanced Real Time Imaging II*, The Minerals,
Metals & Materials Series, https://doi.org/10.1007/978-3-030-06143-2_11

within 0.3 s, and the molten steel is directly contacted with the cooling copper mold, the interfacial heat transfer through the melt–mold interface is extremely important for controlling the solidification of molten materials and solidification structure in these processes, as the heat flux approaches 5–10 MW/m^2 or even higher, which is far higher than in the traditional continues casting (1–2 MW/m^2) [3–5].

Two major types of experimental apparatus have been developed for developing the process of sub-rapid solidification. A number of researchers have found that the heat transfer was affected by many factors [4–10], such as the degree of superheat, the protective atmosphere, the roughness of substrate, and the thickness of the naturally deposited film by using these two apparatuses. Todoroki and Nolli [5–7] developed the droplet solidification technique at Carnegie Mellon University, where the research focused on the heat transfer behavior in strip casting has been conducted. Nolli [6, 8], Yu [9], and Zhu [10], respectively, indicated that sulfur-containing atmospheres, oxygen partial pressure, and coating material could be used to affect heat transfer rates in strip casting. Strezov [11] and Strezov et al. [12] have applied an immersion-type experimental method for investigation of sub-rapid solidification of molten steel, which was designed for modeling the contacting condition between the liquid steel and the roll caster, and they suggested that the heat flux was increasing rapidly due to the natural formed film. Recently, Wang [13] indicated that the heat transfer behavior was greatly affected by the naturally deposited film via using droplet solidification technique. Meanwhile, some scholars have speculated that the heat transfer behavior was also related to the interfacial wettability between the molten steel and copper mold.

In this study, a droplet solidification technique was further developed by Steel Research Center at Central South University. A charge coupled device (CCD) camera is placed beside the bell jar aimed to record the whole melting and solidification process of the steel sample during the dipping test. The effect of interfacial wettability on the heat transfer could be found according to the video data.

Experimental Apparatus and Procedure

Sample Preparation

The composition of the pure Fe steel sample used in the study is listed in Table 1, which was first cut into cylindrical pieces by means of wire electrode cutting with the weight of 5.0 g (±0.1). Then each sample was polished by grinding paper to clean the impurities on the sample surface and subjected to droplet solidification test.

Experimental Procedure

The droplet solidification technique for the in situ observation of the solidification phenomena of metal droplets impinging onto the water-cooled copper substrate was used in this study. The schematic of the system is shown in Fig. 1. First, the sample was placed at the bottom of a quartz tube and melted by the induction coils inside the bell jar. Then the body reaction chamber was evacuated by a vacuum pump and filled up with high purity argon, which has been purified by the atmosphere controlling system. The high purity argon was injected until the time when the oxygen partial pressure of inlet and outlet was consistent, which was 10^{-5} atm; then the sample was melted by the programmed induction furnace. Once the sample got melted, it was ejected onto the copper mold through a small hole at the bottom of the tube by means of a pulse of ultra-high argon from a nozzle installed above the tube. Then, the responding temperatures at different depth (one is 1 mm and the other is 5 mm) were detected by the thermocouples, according to which, the interfacial heat fluxes can be calculated via inverse heat flux algorithm program (IHCP) mathematical model [14–16]. A charge coupled device (CCD) camera is placed beside the bell jar aimed to record the whole melting and solidification process of the steel sample during the dipping tests. The series droplet tests were carried out four times consecutively, and the copper mold was not cleaned during experiments in order to deposit the natural oxide film and investigate the effect on the corresponding heat fluxes.

Table 1 Composition of the industrial pure Fe steel (mass%)

	C	Si	Mn	P	S	Cr	Ni	Fe
Pure Fe	0.003	0.02	0.14	0.012	0.010	0.005	0.02	Bal.

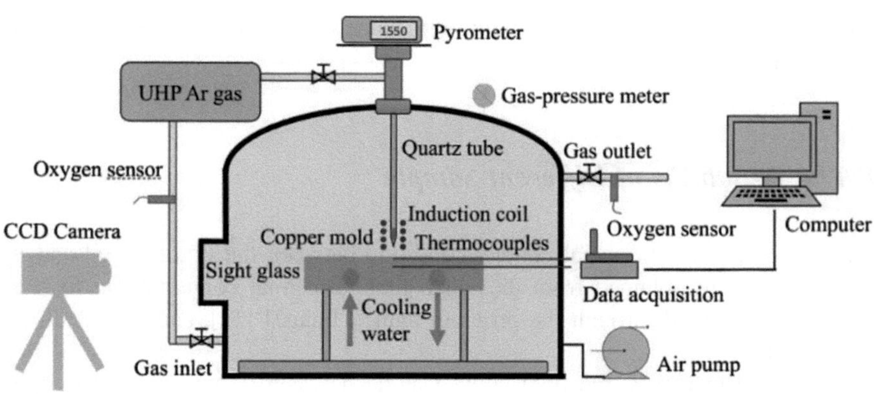

Fig. 1 Schematic of droplet solidification technique

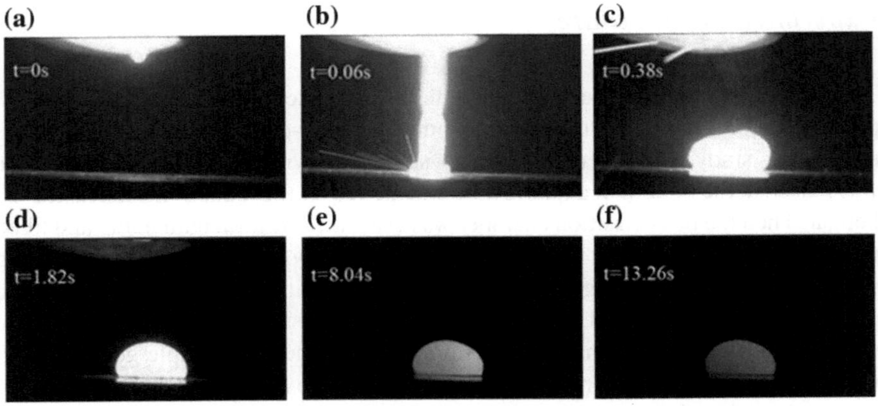

Fig. 2 Whole process of ejection and solidification of droplet solidification test (**a**) the start of ejection; (**b**) the beginning of the contact between liquid steel and copper mold; (**c**) the forming of the liquid steel; (**d–f**) the process of solidification

Results and Discussion

A The Solidification Process of the Steel Sample

The whole process of the droplet solidification was captured by the CCD camera. An example ejection and solidification of the droplet was given in Fig. 2. As shown in Fig. 2, the sample was ejected from the quartz tube first (Fig. 2a). Then, the sample was in contact with the substrate at 0.06 s (Fig. 2b). From 0.06 to 1.82 s, the sample was forming the hemisphere shape on the substrate (Fig. 2c, d). After 1.82 s, the sample was solidified gradually, while the shape of the sample was not changed (Fig. 2e, f).

B The Wettability of Different Sample

The photos of the interfacial contact condition between the liquid metal and the copper mold are obtained from the video files. The final contact angle for each sample is measured through the program named ImageJ [17], and the results are shown in Fig. 3.

As shown in Fig. 3, the final contact angle between liquid metal and copper substrate is 140.931°, 129.748°, 117.273°, 95.112° for the first, second, third, and fourth sample, respectively.

The relationship between the contact angle and surface tension could be described by Young's equation, which is expressed as Eq. (1).

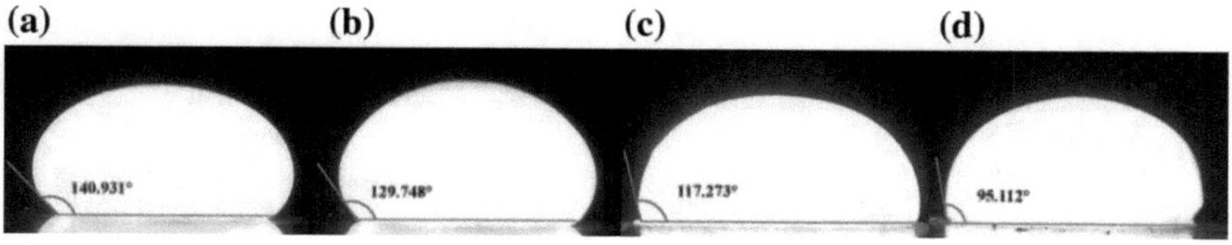

Fig. 3 Variation of the final contact angle of different sample: **a** first, **b** second, **c** third, and **d** fourth sample

$$\gamma_{s/g} = \gamma_{l/g} + \gamma_{l/s} \cos\theta \quad (1)$$

In this formula, $\gamma_{s/g}$, $\gamma_{l/g,}$ and $\gamma_{l/s}$ represent the surface tension of the substrate, the liquid sample, and the interface tension between liquid and substrate, respectively. θ is the contact angle.

Considering that the atmosphere and the steel sample are the same for each individual test, $\gamma_{s/g}$ and $\gamma_{l/g}$ are assuming constant in this study. Then, the contact angle is influenced mainly by interfacial surface tension of steel/substrate, $\gamma_{s/l}$. Observing the final contact angle can indirectly illustrate the adhesion of the sample and the substrate. When the contact angle decreases, better wetting condition is formed between the sample and the substrate. This will accelerate the filling of molten steel into the roughness cavities, thus the air gap between the steel sample and the copper mold is reduced, leading to the reduction of interfacial thermal resistance.

C The Heat Transfer of Different Sample

In order to study the influence of interfacial wettability between the copper mold and the sample on the interfacial heat transfer behavior, the calculated heat fluxes and amounts of heat removed during the first 3 s are plotted versus time, as shown in Fig. 4a, b. As expected, the heat fluxes in all experiments show a similar variation pattern for pure Fe steel samples, in which the heat fluxes pick rapidly within first 0.4 s during the initial contact of the metal with the substrate and then decrease gradually till a final relatively stable state. The reduction of the heat fluxes after the maximum value could be explained as the reduction of the latent heat dissipation. Also, it may be associated with the increase of the overall thermal resistance between the solidified droplet and mold, which is caused by the contraction of the solidified layer and consequent formation of an air gap between the substrate surface and the solidified shell.

In the heat removed plots (Fig. 4b), the curves are grouped in bundles: The steeper a bundle is, the better the heat transfer conditions are for that condition [5, 8]. The heat transfer condition in different experiment is quite different from each other.

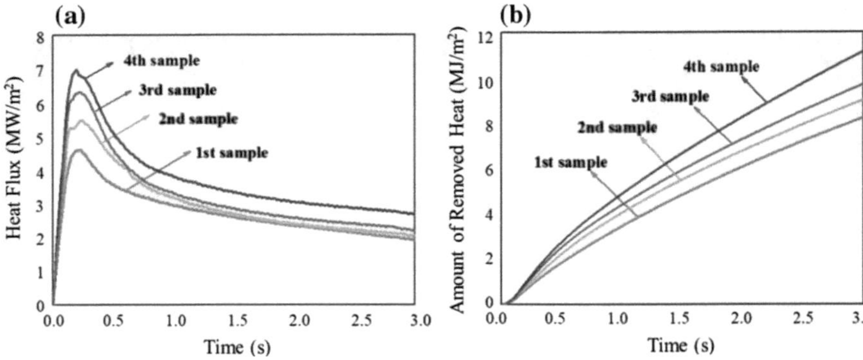

Fig. 4 **a** Variation of interfacial heat flux with time and **b** variation of amount of removed heat with time

Fig. 5 Variation of peak heat flux and amount of removed heat at first 3 s of different samples

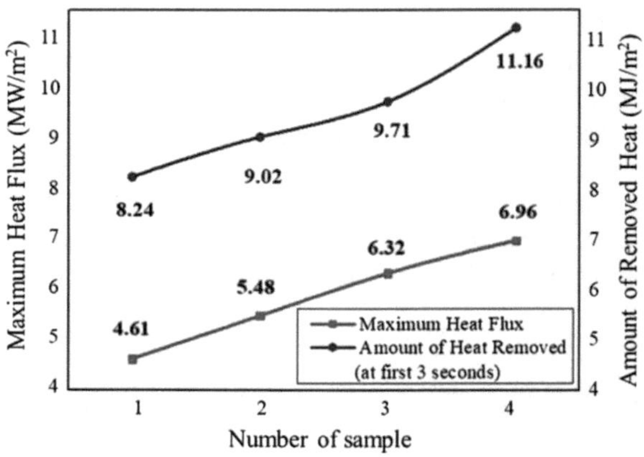

Figure 5 shows the maximum heat flux values and the amount of heat removed per unit surface area during the first 3 s versus the number of the sample. It can be clearly observed that both maximum values for heat flux and the amount of heat removed show a similar variation trend that increases due to better wetting condition.

The heat flux during the droplet test, as the controlling factor for casting yield and solidification structure, is affected by many factors, such as the degree of superheat, the protective atmosphere, the roughness of substrate, and interfacial wetting conditions [5, 8–11], in which the dominate one is the interfacial thermal resistance. The improved wettability of the substrate and the liquid steel suggests a possible explanation for the reason why the general heat flux observed under different wetting condition is increased. With the reduction of the final contact angle, the melt–substrate interface energies between liquid metal and copper mold have been lowered and made it easier for the liquid metal to fill the surface roughness cavities and generate smaller air gap; meanwhile, the real interfacial contact area increases, and thus the general heat transfer rate is improved with the decreasing of interfacial resistance.

Conclusion

The droplet solidification technique has been further developed to study the effect of interfacial wettability on heat transfer. The main conclusions are listed as follows:

1. The developed droplet solidification technique could clearly observe the whole process of the droplet solidification. The final contact angle between the sample and the substrate could be determined according to the video file.
2. The effect of interfacial wettability between the steel sample and the copper mold in the interfacial heat transfer is obvious. The heat flux for casting pure Fe steel is increasing, due to a better interfacial contact condition, which tends to improve the contact area and reduce the interfacial thermal resistance.

References

1. Netto PGQ, Tavares RP, Isac M, Guthrie RIL (2001) A technique for the evaluation of instantaneous heat fluxes for the horizontal strip casting of aluminum alloys. ISIJ Int 41(11):1340–1349
2. Luiten EEM, Blok K (2003) Stimulating R&D of industrial energy-efficient technology; the effect of government intervention on the development of strip casting technology. Energy Policy 31(13):1339–1356
3. Loulou T, Artyukhin EA, Bardon JP (1999) Estimation of thermal contact resistance during the first stages of metal solidification process: I—experiment principle and modelisation. Int J Heat Mass Tran 42(12):2129–2142
4. Zhang W, Yu Y, Fang Y, Li J (2011) Determination of interfacial heat flux of stainless steel solidification on copper substrate during the first 0.2 s. J Shanghai Jiaotong Univ 16(1):65–70
5. Nolli P (2007) Doctoral thesis. Carnegie Mellon University
6. Todoroki H, Lertarom R, Cramb AW, Jimbo I, Suzuki T (1996) Evaluation of the initiation of solidification of iron against a water cooled copper mold. Electr Furn Conf Proc, 1371–1379
7. Nolli P, Cramb AW (2008) Naturally deposited oxide films in near-net-shape casting: importance, mechanisms of formation, and prediction. Metall Mater Trans B 39(B):56–65
8. Nolli P, Cramb AW (2007) Interaction between iron droplets and H_2S during solidification. ISIJ Int 47:1284–1293
9. Yu Y, Cramb AW, Heard R, Fang Y, Cui J (2006) The effect of oxygen partial pressure on heat transfer and solidification. ISIJ Int 46(10):1427–1431
10. Zhu CY, Wang WL, Lu C (2019) Characterization of cermet coatings and its effect on the responding heat transfer performance in strip casting process. J Alloy Compd 770:631–639
11. Strezov L, Herbertson J (1998) Experimental studies of interfacial heat transfer and initial solidification pertinent to strip casting. ISIJ Int 38:959–966
12. Strezov L, Herbertson J, Belton GR (2000) Mechanisms of initial melt/substrate heat transfer pertinent to strip casting. Metall Mater Trans B 31(B):1023–1030
13. Wang WL, Zhu CY, Lu C, Yu J, Zhou LJ (2018) Study of the heat transfer behavior and naturally deposited films in strip casting by using droplet solidification technique. Metall Mater Trans A
14. Zhang H, Wang W, Zhou D, Ma F, Lu B, Zhou L (2014) A study for initial solidification of Sn–Pb alloy during continuous casting: Part I. The development of the technique. Metall Mater Trans B 45(B):1038–1047
15. Zhou D, Wang W, Zhang H, Ma F, Chen K, Zhou L (2014) A study for initial solidification of Sn–Pb alloy during continuous casting: Part II. Effects of casting parameters on initial solidification and shell surface. Metall Mater Trans B 45(B):1048–1056

16. Zhang H, Wang W (2017) Mold simulator study of heat transfer phenomenon during the initial solidification in continuous casting mold. Metall Mater Trans B 48(B):779–793
17. Luo X, Wang W, Ma F (2016) Degree of undercooling and wettability behavior of liquid steel on single-crystal Al_2O_3 and MgO substrate under controlled oxygen partial pressure. ISIJ Int 56(8):1333–1341

Comparison of Dissolution Kinetics of Nonmetallic Inclusions in Steelmaking Slag

Mukesh Sharma and Neslihan Dogan

Abstract Nonmetallic oxide inclusions of Al_2O_3, Al_2TiO_5, and $CaO \cdot 2Al_2O_3$ (CA2) types are responsible for clogging of ceramic nozzles during liquid steel processing. The dissolution of these inclusions in steelmaking slags alleviates the clogging phenomenon. The in situ dissolution behavior a single oxide particle is studied in a synthetic $CaO–Al_2O_3–SiO_2$ type slag using a high-temperature confocal scanning laser microscope at 1550 °C. The rate determining step for Al_2O_3 and CA2 inclusions was confirmed to be mass transport control in slag. The rate determining step for dissolution of Al_2TiO_5 needs further investigation. The rate of dissolution varied in the order from slowest to fastest: $Al_2O_3 < CA2 < Al_2TiO_5$.

Keywords Inclusion dissolution · Alumina · Aluminum titanate
Calcium aluminate · Confocal scanning laser microscope

Introduction

The formation of nonmetallic inclusions is inevitable during liquid steel refining. These inclusions cause nozzle clogging during continuous casting [1–3] and are deleterious to the final properties of the cast product [4]. Therefore, the chemical composition, size, morphology, and distributions of the inclusions should be well controlled. To achieve such control, the inclusions are either modified or removed. The removal of inclusions in reactors such as ladle, tundish, or the mold takes place in three important steps (a) transportation of the inclusion from the metal phase to the slag/metal interface, (b) separation of the inclusion from the slag/metal interface to the slag, and (c) dissolution of the inclusion in slag [5–11]. The discussion in this paper focuses on the studies related to the third step, i.e., dissolution of inclusion in slag bulk. Regarding the topic of inclusion dissolution in the slag, the high-temperature confocal scanning laser microscope (CSLM) is a relatively novel methodology to be applied in this field. The pioneer work is dated back to Emi's group [12]. Over the years, the CSLM has been applied for the observations of col-

M. Sharma (✉) · N. Dogan
Steel Research Center, McMaster University, Hamilton, ON L8S 4L7, Canada
e-mail: sharmmk2@mcmaster.ca

J. Nakano et al. (eds.), *Advanced Real Time Imaging II*, The Minerals, Metals & Materials Series, https://doi.org/10.1007/978-3-030-06143-2_12

lision, agglomeration, and cluster formation behaviors of inclusions in steel/Ar gas and the steel/slag interfaces [4–10], the phenomena of solidification [13, 14] and phase transformations [15–17] in steel. Majority of studies related to the dissolution of oxide inclusions using CSLM focuses on Al_2O_3 inclusion particles [5–7, 11, 18–21]. Few studies investigate other oxides such as MgO [20, 22–24], $MgAl_2O_4$ [22, 25, 26], SiO_2, [27], and CaO [28]. Recent studies at McMaster University has focused on dissolution of complex oxide inclusion types such as $CaO \cdot 2Al_2O_3$ [29] and Al_2TiO_5 [30] along with TiO_2 [31]. This work compares the dissolution rate of Al_2O_3, $CaO \cdot 2Al_2O_3$ (CA2) and Al_2TiO_5 particles using CSLM under conditions relevant to steelmaking operations.

Experimental Methodology

The key features of CSLM include (1) a confocal imaging system with a laser which provides sharp image with high contrast and resolution and (2) an infrared heated gold image furnace which provides fast heating and cooling rates at steelmaking temperatures with an excellent temperature control. The temperature is measured using a B-type thermocouple. The CSLM crucible and sample configuration for Al_2TiO_5 dissolution is published elsewhere [32]. The particle and platinum crucible containing pre-fused slag are heated to the experimental temperature in a 99.999% pure argon atmosphere. The predetermined heating and cooling schedule for dissolution experiments is as follows: heating from room temperature to 150 °C at the rate of 40 °C/min, followed by heating at the rate of 1000 °C/min to a temperature which is 100 °C below the desired experimental temperature for Al_2TiO_5 particles, whereas a heating rate of 150 °C/min is used for Al_2O_3 and CA2 particles. Thereafter the temperature is raised to the dissolution temperature at the rate of 100°C/min. It is held at the experimental temperature until dissolution is completed, and the system is then cooled to room temperature at the rate of 100 °C/min. A faster heating rate is used in case of Al_2TiO_5 particle dissolution experiments because they dissolve much faster than Al_2O_3 or CA2 particles. The method for preparation and characterization of the slag is published elsewhere [30, 32]. A $23.2Al_2O_3$–30.5CaO–$46.3SiO_2$, all in weight percent, type slag is used in the current work for studying dissolution. The slag liquidus is 1442 °C which is calculated using FactSage 7.2© Equilib program. The density of slag was calculated from model provided in Slag Atlas [33]. The Al_2O_3 particles are purchased from Alfa Aesar, USA, in the form of irregular particles of 1–3 mm. The CA2 particles are obtained by sintering the Al_2O_3 and CaO powders, the details of which are published by Miao et al. [29]. The Al_2TiO_5 particles are obtained from crystals prepared by the optical floating zone method (OFZM). The details of OFZM are published in a separate publication [31]. All the particle types are crushed to obtain particles approximately 300–900 μm of equivalent circular diameter. The crushing process creates a problem of irregular particle shape which affects the kinetic treatment of data, discussed later in the paper.

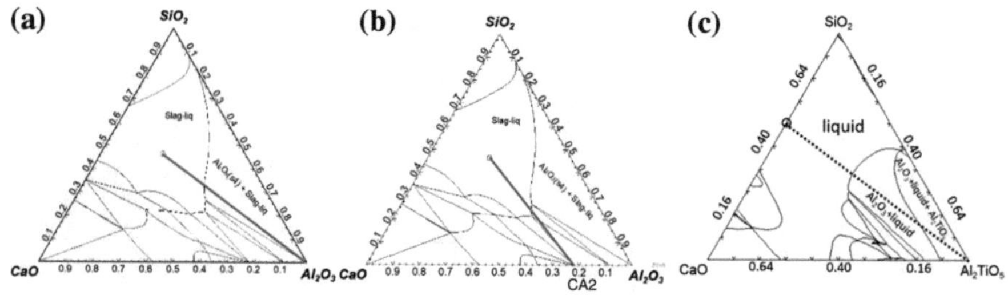

Fig. 1 Phase diagram for the **a, b** CaO–SiO_2–Al_2O_3 system and **c** CaO–SiO_2–A12O3–Al_2TiO_5 system, showing the studied slag system (hollow circle) assuming straight dissolution path (straight line)

Thermodynamics of Inclusion Dissolution

As the dissolution proceeds, various phases may form and disappear dynamically at the interface of the particle and the slag. An isothermal equilibrium phase diagram can help identify the stable phases that form at the interface. The phase diagrams shown in Fig. 1 are constructed using Factsage 7.2^{TM}. For Al_2O_3 and CA2 dissolution, the phase that might form at the interface is $CaAl_2Si_2O_8$ whereas it is Al_2O_3 for Al_2TiO_5 dissolution, Al_2O_3 is a stable phase which might form at the interface.

Results and Discussion

In Situ Observation of Inclusion Dissolution

The in situ dissolution of particles in slag using CSLM is recorded in the form of a video. Frames from the video are extracted to measure the size of particle with time to delineate the dissolution kinetics. The particle is identified by drawing a boundary around the particle, shown in Fig. 2. The area inside the boundary is measured, and this area is equated to the equivalent area of a circle. The equivalent circular radius is used conventionally to show the size change with respect to time. Figure 2 shows the images of Al_2O_3, CA2, and Al_2TiO_5 type of particles during in situ dissolution in CSLM. The Al_2O_3 and CA2 particles are transparent to CSLM while the Al_2TiO_5 particles are not. The difference in hue of the Al_2TiO_5 particle is most likely due to the presence of Ti, a transition element. The slag does not consist of any transition element and is therefore transparent to the laser. Moreover, the particles show different rotation behavior. The Al_2O_3 particles rotate from the beginning to the end of dissolution. The Al_2TiO_5 particles do not rotate until the very end of dissolution. They might rotate occasionally. The CA2 particles do not show any rotation during dissolution.

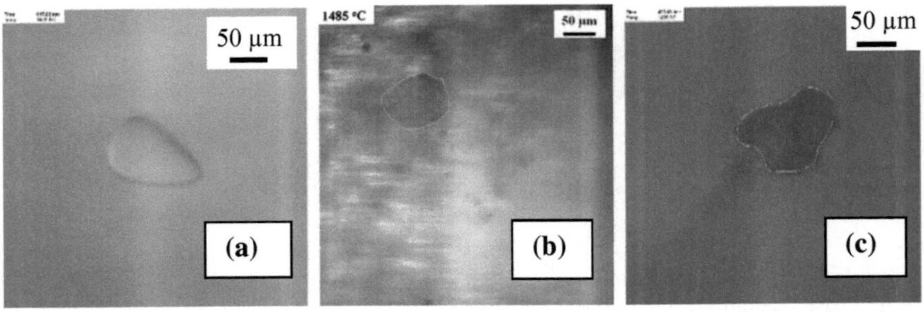

Fig. 2 Images of **a** Al_2O_3, **b** CA2, and **c** Al_2TiO_5 particles during in situ dissolution in slag

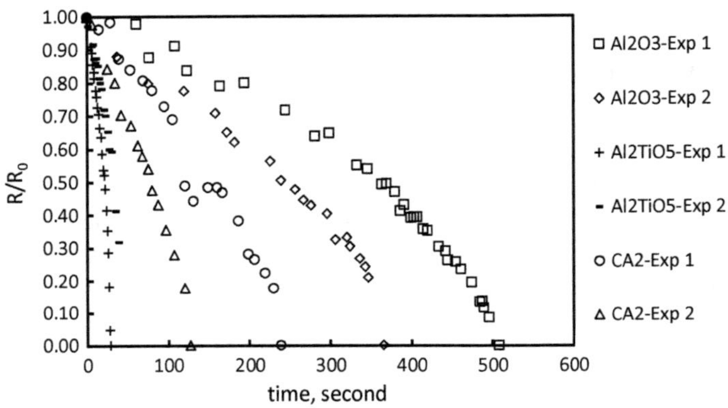

Fig. 3 Changes in normalized equivalent radius of Al_2O_3, Al_2TiO_5, and CA2 particles with time at 1550 °C

Dissolution Rate

Figure 3 compares the changes in the normalized equivalent circular radius of Al_2O_3, CA2, and Al_2TiO_5 particles with time at 1550 °C. There are two experimental data presented here for each case. For CA2 particles of diameter of 250 and 230 μm, the dissolution time are 238 and 127 s, respectively. For Al_2O_3 particles of diameter of 318 and 314 μm, the dissolution time are 508 and 365 s, respectively. For Al_2TiO_5 particles of diameter of 402 and 902 μm, the dissolution time are 29 and 39 s, respectively. It can be easily inferred that the rate of dissolution is highest for Al_2TiO_5 particle, followed by CA2, followed by Al_2O_3. The reasons for this observation are discussed in the next section.

Dissolution Mechanism

In previous work by the authors [29–32], the dissolution of alumina, calcium aluminate, and aluminum titanate particles is found to be mass transfer controlled. Assuming mass transfer control in slag, the shrinking core model (SCM) can be used to estimate the diffusivity of the rate determining species. For the dissolution of Al_2O_3, CA2, and Al_2TiO_5 particles, the rate determining species can be assumed to be Al–O anion. The SCM for spherical particle with reducing size, shown in Eq. 1, is used to estimate the diffusivity of alumina for dissolution of different oxides [34].

$$\left(\frac{R}{R_0}\right)^2 - 1 = \frac{2b\Delta C D}{\rho_{\text{particle}} R_0^2}t \tag{1}$$

where R and R_0 are instantaneous and initial radius in m, t is time in s, ρ_B is the density of particle in kg/m^3, b is a stoichiometric constant and equal to 1. ΔC is the difference in concentration of the alumina in bulk slag and slag at saturation, and D is the diffusivity of alumina. The ΔC values for Al_2O_3, CA2, and Al_2TiO_5 particles were calculated using Factsage^{TM} thermochemical software program. The ΔC values are 620, 757, and 301 kg/m^3 for Al_2O_3, CA2, and Al_2TiO_5 particles, respectively. The slope of the $(R/R_0)^2-1$ versus t line can be obtained from the experimental data. The value of D is obtained by placing all the known parameters as shown in Eq. 1. These values for each experiment are summarized in Table 1.

In Fig. 4, the CSLM data for dissolution of Al_2O_3, CA2, and Al_2TiO_5 particles is compared against the model represented by Eq. 1. The CSLM data is analyzed with respect to the regression values obtained by fitting curve from the SCM. This approach is adapted from the study by Monaghan and Chen [25]. This method is useful in determining the degree of linearity of the kinetic data with respect to dissolution mechanism. The linearity is quantified in terms of the degree of fitting parameter R^2. The values of R^2 for Al_2O_3 are 0.99 and 0.95 while that for CA2 is 0.95 and that for Al_2TiO_5 is close to 1. All the values are between 0.9 and 1, and this implies that the dissolution process is most likely controlled by the mass transfer in the slag.

The estimated diffusivity of Al_2O_3 and CA2 particles is in the order of 10^{-10} m^2/s. These values are comparable to the diffusivity values reported in the literature [33]. Under the assumption of mass transfer control in slag, the dissolution kinetics

Table 1 Estimated diffusivity values of Al_2O_3 assuming mass transfer control

Particle	Al_2O_3 Exp-1	Al_2O_3 Exp-2	Al_2TiO_5 Exp-1	Al_2TiO_5 Exp-2	CA2 Exp-1	CA2 Exp-2
Slope (Fig. 4)	−0.0021	−0.0029	−0.035	−0.022	−0.0047	−0.0089
R_0, m	1.6E−04	1.6E−04	2.0E−04	4.5E−04	1.3E−04	1.1.2E−04
D Al_2O_3, m^2/s	1.6E−10	2.2E−10	89E−10	280E−10	1.8E−10	3.0E−10

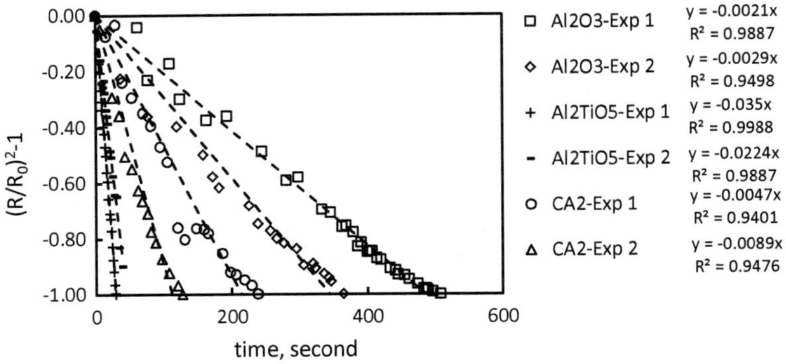

Fig. 4 Comparison of obtained CSLM data for Al_2O_3, Al_2TiO_5, and CA2 with that of SCM for particles with decreasing size

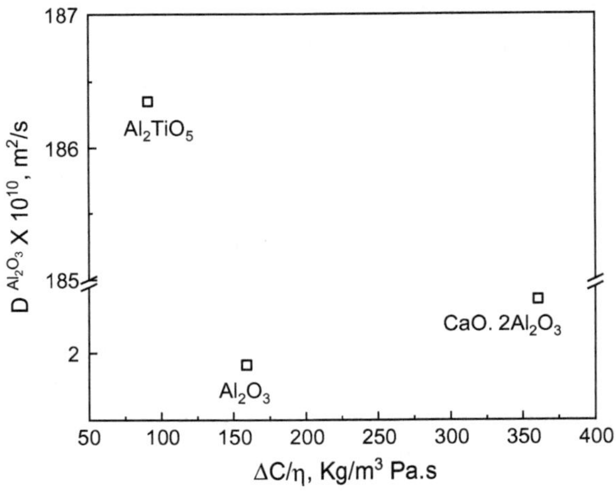

Fig. 5 Diffusivity compared to viscosity of saturated slag for Al_2O_3, Al_2TiO_5, and CA2 particles

should theoretically follow the same order as that of ΔC. The value of ΔC for CA2 particles is 757 kg/m^3 which is greater than that for Al_2O_3 particles (620 kg/m^3). The experimental finding is valid for the mass transfer control assumption. Also, Eyring's relationship suggests that diffusivity should be inversely proportional to viscosity of slag [33]. Figure 5 shows a comparison of the obtained diffusivity values against a ratio of $\Delta C/\eta$. Here η is the viscosity of the slag saturated with the particle and is estimated from Urbain's model [33]. It is found that the alumina saturated slag is more viscous than CA2 saturated slag. This indicates that dissolution of alumina particles is slower than CA2 particle and this is also observed from CSLM data. Thus, it can be confidently inferred that the dissolution of Al_2O_3 and CA2 particles is mass transfer control in slag.

From Table 1, the diffusivity values obtained for Al_2TiO_5 particles is in the range of 10^{-8} m^2/s which is two order of magnitude higher than those for other particles in a similar slag composition [33]. From experimental observations shown in Fig. 3, the aluminum titanate particle dissolves much faster than the Al_2O_3 or the CA2 particles. Considering thermodynamic factors of ΔC and η, the $\Delta C/\eta$ for Al_2TiO_5 is the lowest as shown in Fig. 5. Thus, if dissolution of Al_2TiO_5 particle is controlled by mass transfer in slag then the dissolution of Al_2TiO_5 should be slowest. The observations do not support the theory for this particle. Thus, it can be argued that the transport mechanism of aluminum titanate is different than the other two particles. It is known that Al_2TiO_5 is unstable at low oxygen potential, dissociating into Al_2O_3, Ti_2O_3, and O_2 [35]. The oxygen gas bubbles form and burst within the liquid slag as observed in the CSLM experiments. There is enhanced localized mass transport in the slag due to bursting of gas bubbles. The effect of stirring is not included in the diffusivity obtained from Eq. 1 which assumes a Sherwood number of 2. Yan et al. [36] modified the mass transfer coefficient for dissolution of alumina to include the effect of rotation on stirring. A similar modification to the mass transfer coefficient is suggested to study the effect of gas stirring due to oxygen. Further the importance of dissociation reaction on the overall reaction kinetics should be studied.

Effect of Particle Shape on Dissolution Rate

The particles used in this work are obtained by crushing larger irregular particles; the shape of the particle during experiment is uncontrolled and irregular. From Fig. 2, it is observed that the particles are mostly non-spherical in shape. However, when the kinetic analysis is done using Eq. 1 based on a SCM, shown in Figs. 3 and 4, the particle is assumed to be sphere for the sake of simplicity. The effect of particle shape is studied by comparing the rate of change in size of irregular particles with respect to the rate of change in size of spherical particles. The three sets of spherical data used, named 'spherical,' in the legend are obtained from work by Liu et al. [7], shown in Fig. 6. The experimental data for irregular-1 and irregular-2 in Fig. 5 for alumina particles were obtained from the current study. The average dR/dt for Spherical-1, Spherical-2, and Spherical-3 case is 0.20, 0.19, and 0.13 μm/s while the average dR/dt for irregular-1 and irregular-2 is 0.43 and 0.58 μm/s. In the case of dissolution of spherical particles, the rotation of these particles is either negligible or does not exist. Therefore, the rate of change in particle size is relatively steady. The average rate of change in size for spherical alumina particles is 2–3 times slower than the average rate of irregular alumina particles. This is most likely due to the rotation which causes local convection and enhances the rate of transfer of mass from the particle to the slag. Thus, controlled spherical shape of particles is important in proper kinetic analysis of CSLM data.

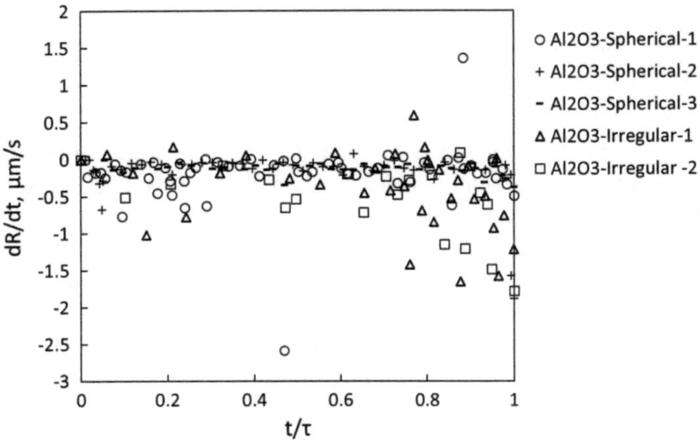

Fig. 6 Comparison of rate of change in size of regular and irregular alumina particle

Conclusion

The dissolution of microparticles of Al_2O_3, CA2, and Al_2TiO_5 in a $CaO–SiO_2–Al_2O_3$ type slag at 1550 °C using CSLM is presented. The dissolution mechanism is found to be mass transport control in slag for Al_2O_3 and CA2 inclusions. The dissolution mechanism for Al_2TiO_5 particles is more complex and clearly evades the assumption for mass transfer control. The effect of gas generation on the dissolution kinetics should be further analyzed. For the oxides discussed, the inclusions are ranked in order of increasing dissolution rates as follows: $Al_2TiO_5 > CA2 > Al_2O_3$.

References

1. Rackers KG, Thomas BG (1995) 78th steelmak. Conf Proc 78:723
2. Zhang L, Thomas BG (Nov. 2003) XXIV National Steelmaking Symposium. Morelia, Mich, Mexico 26–28:138–183
3. Cramb AW, Maddalena RL (2003) Refract materials, 11th edn. AISE Steel Foundation, Pittsburgh, pp 2–5
4. Thornton PA (1971) J Mater Sci 6:347
5. Lee SH, Tse C, Yi KW, Misra P, Chevrier V, Orrling C, Sridhar S, Cramb AW (2001) J Non Cryst Solids 282:41
6. Valdez M, Prapakorn K, Cramb AW, Sridhar S (2002) Ironnmak Steelmak 29:47
7. Liu J, Verhaeghe F, Guo M, Blanpain B, Wollants P (2007) J Am Ceram Soc 90:3818
8. Sridhar S, Cramb AW (2003) High Temp Mater Process 22:275
9. Valdez M, Shannon GS, Sridhar S (2006) ISIJ Int 46:450
10. Michelic S, Goriupp J, Feichtinger S, Kang YB, Bernhard C, Schenk J (2016) Steel Res Int 87:57
11. Monaghan BJ, Chen L (2004) J Non Cryst Solids 347:254
12. Yin HB, Shibata H, Emi T, Suzuki M (1997) ISIJ Int 37:936
13. Orrling C, Phinichka N, Sridhar S, Cramb AW, Fang Y (2000) JOM 51:1

14. Chikama H, Shibata H, Emi T, Suzuki M (1996) Mater Trans 37:620
15. Shibata H, Arai Y, Suzuki M, Emi T (2000) Metall Mater Trans B 31:981
16. Phelan D, Reid MH, Dippenaar R (2005) Microsc Microanal 11:670
17. Sridhar S (2005) Application of confocal scanning laser microscopy to steel research. In: The 3rd international congress on the science and technology of steelmaking, Charlotte, NC. AIST, Warrendale, PA
18. Verhaeghe F, Liu J, Guo M, Arnout S, Blanpain B, Wollants P (2007) Appl Phys Lett 91:124104
19. Sridhar S, Cramb AW (2000) Metall Mater Trans B 31:406
20. Yi KW, Tse C, Park JH, Valdez M, Cramb AW, Sridhar S (2003) Scand J Metall 32:177
21. Monaghan BJ, Chen L (2005) Steel Res Int 76:346
22. Fox AB, Valdez ME, Gisby J, Atwood RC, Lee PD, Sridhar S (2004) ISIJ Int 44:836
23. Park JH, Jung I, Lee H (2006) ISIJ Int 46:1626
24. Liu J, Guo M, Jones PT, Verhaeghe F, Blanpain B, Wollants P (2007) J Eur Ceram Soc 27:1961
25. Monaghan BJ, Chen L (2006) Ironmak Steelmak 33:323
26. Monaghan BJ, Chen L, Sorbe J (2005) Ironmak Steelmak 32:258
27. Feichtinger S, Michelic SK, Kang YB, Bernhard C (2014) J Am Ceram Soc 97:316
28. Guo X, Guo M, Sun Z, Van Dyck J, Blanpain B, Wollants P (2010) Materials science and technology conference, pp 1739–1750
29. Miao K, Haas A, Sharma M, Mu W, Dogan N (2018) Metall Mater Trans B 49:1612
30. Sharma M, Mu W, Dogan N (2018) JOM 70:1220
31. Sharma M, Dabkowska HA, Dogan N (2018) Steel Res Int 1800367:1
32. Sharma M, Mu W, Dogan N (2018) AISTech 2018, pp 2601–2608
33. Allibert M, Gaye H, Geisler J, Janke D, Keene BJ, Kirner D, Kowalski M, Lehmann J, Mills KC, Neuschutz D, Parra R, Saint Jours C, Spencer PJ, Susa M, Tmar M, Woermann E (1995) Slag Atlas, 2nd edn. Dusseldorf, p 48
34. Levenspiel O (1999) Chemical reaction engineering, 3rd edn. Wiley, New York
35. De Arenas IB (2012) In: Lakshmanan DA (eds) Sintering of ceramics—new emerging techniques, 1st edn. InTech, p 503
36. Yan P, Webler BA, Pistorius PC, Fruehan RJ (2015) Metall Mater Trans B 46:2414

Part VII
Phase Transformation I

Quantitative Thermal Analysis of Solidification in a High-Temperature Laser-Scanning Confocal Microscope

Dasith Liyanage, Suk-Chun Moon, Madeleine Du Toit and Rian Dippenaar

Abstract Under near-equilibrium conditions, the concentric solidification technique proved to be an excellent way of studying in situ, solidification phenomena, but under rapid cooling conditions, the solid/liquid interface undergoes dynamic thermal and solute distributions. The current project aims to evaluate temperature distribution under rapid cooling conditions. A number of thermocouples are attached to the specimen surface and measured the temperature over the solidification period. The temperature profile within the liquid phase is measured separately by thinner thermocouple wires which are fixed to the crucible so that the surface tension of the molten liquid keeps the thermocouple suspended in the liquid pool. The dynamically changing temperature profile over the radial axis of the specimen under rapid cooling conditions is determined, including the all-important temperature at the solid/liquid interface. The calculated interface temperatures are utilized in phase-field simulations, and the results are found to be in excellent agreement with experimental results.

Keywords HTLSCM · Phase transformation · Interface kinetics
Transient cooling

Introduction

In a series of pioneering experiments, Professor Toshi Emi [1] and his group in Tohoku University demonstrated for the first time how the unique features of high-temperature laser-scanning confocal microscopy can be utilized to study solidification events in metallic systems. A specimen is heated by radiation in an infrared furnace and images are created by scanning a He–Ne laser beam, wavelength 632.8 nm, two dimensionally (15.7 kHz × 60 Hz) across the specimen. The images are recorded in real time at a rate of 30 fps [1, 2]. A thin optical plane is brought into

---

D. Liyanage (✉) · S.-C. Moon · M. Du Toit · R. Dippenaar
School of Mechanical, Materials, Mechatronic and Biomedical Engineering,
University of Wollongong, Wollongong, NSW 2522, Australia
e-mail: dddl983@uowmail.edu.au

J. Nakano et al. (eds.), *Advanced Real Time Imaging II*, The Minerals,
Metals & Materials Series, https://doi.org/10.1007/978-3-030-06143-2_13

focus by the use of a confocal pinhole, eliminating reflected rays not coming from this plane and most importantly, screening out infrared radiation. A high-resolution image is produced at high temperature, but a consequence of creating a thin optical plane is that it is difficult to conduct solidification studies where a curved surface, such as a meniscus forms. In order to overcome this impediment, Mark Reid and Dippenaar [2] developed the so-called concentric solidification technique. Because there is an inverted V-shaped temperature profile in the furnace, a stable liquid pool is created in a thin, cylindrical disk, which is supported by a solid rim. The rate and morphology of the solid/liquid interface can then be tracked on cooling in real time and at temperature.

Shibata et al. [3] and Arai et al. [4] were the first to utilize laser-scanning confocal microscopy to study the peritectic phase transition in Fe–C and Fe–Ni alloys, respectively. They verified experimentally, earlier suggestions by, e.g., Kerr et al. [5] and Stefanescu [6] that the peritectic reaction occurs by the growth of a thin austenite layer on the liquid/delta ferrite interface (peritectic reaction) followed by the growth of this austenite layer into the liquid and back into solid, respectively (the peritectic transformations).

The excitement evoked by a fundamental study of the intriguing peritectic phase transition might best be illustrated by a remark found in the introduction to Stefan Griesser's Ph.D. thesis [7] "*The peritectic transition is arguably the most interesting topic in the world since it involves birth, competition and growth of a new intermediate phase between two interacting parent phases of diverse physical properties, ruled by kinetic and thermodynamic laws.*" Moreover, the peritectic phase transition in Fe–C alloys is not only of academic interest, but it is also of significant importance to the steel industry, since steel of near-peritectic composition is notoriously difficult to cast and many of the global steel produced fall into this composition range.

Shibata et al. [3] found that the growth of the newly nucleated phase in hyper-peritectic Fe–C alloys can be satisfactorily explained by a model based on diffusion, but the high rate of the peritectic transformation in hypo-peritectic cannot be explained by such a model. Emi and his co-workers [1] estimated that the temperature gradient at the solid/liquid interface is about 5 K/mm at very low cooling rates as opposed to theoretical model predictions (Trivedi [8] and Kurz and Fisher [9] of about 1 K/mm). Griesser [7] provided convincing experimental evidence that nucleation events are constrained at high cooling rates by the solute diffusional field surrounding an embryonic nucleus, which leads to substantial undercooling ($\Delta T > 40$ K). Constrain nucleation leads to the occurrence of a massive type of phase transformation [7]. The rate of nucleation is exponentially dependent upon the interface temperature [10] and it is this need to know the exact interfacial temperature that prompted the present study, which forms part of a larger program aimed at quantifying the events occurring in the course of a peritectic phase transition. In addition, such knowledge will be used to validate theoretical models based on the temperature gradient and local undercooling at the solid/liquid interface [11, 12].

Yin and Emi [13] were the first to determine the overall thermal gradient in a solid specimen, 4.3 mm in diameter, by attaching thermocouples to the center and to the outer edge of the sample and determined t gradient as 8 K/mm at a temperature

75 K below the melting point. However, they were unable to measure the temperature of the liquid pool since the thermocouples detached from the surface. Phelan [14] measured experimentally the temperature gradients at different temperatures by attaching three thermocouples to a specimen holder of 12 mm × 5 mm rectangular crucible and found temperature gradients of 3 K/mm and 16 K/mm at temperatures of 823 K and 1823 K, respectively. They argued that this increase in thermal gradient with temperature compounds the difficulty of defining the temperature distribution across the specimen, thereby inhibiting a quantitative analysis of the morphology of solid/liquid interfaces [14].

Mark Reid and Dippenaar [2] were the first to estimate the thermal gradient in a concentric solidification experiment. They placed pure silver and nickel particles in a 9.7-mm-diameter alumina crucible and observed where melting occurred with respect to the center of the crucible and determined temperature gradients of 5 K/mm and 20 K/mm at temperatures of 1234 K and 1723 K, respectively. Griesser [7] estimated the temperature distribution in an Fe–C alloy by measuring experimentally the respective fractions of the solid and liquid phases by following step by step, the progression of the solid/liquid interface and then matching them with the equilibrium phase diagram. They estimated the temperature gradient as 9.8 K/mm at a cooling rate of 2 K/min.

Although Griesser's technique provided valuable information, there is still an urgent need to validate his findings experimentally, especially at cooling rates higher than 1–2 K/min, in addition to determining the temperature distribution over the entire period of solidification instead of just one or two temperatures. The temperature gradient at the solid/liquid interface under dynamic solidification processing is the most important parameter. It also has to be borne in mind that an additional thermal resistance may be present at this interface because of the different thermal conductivities of the melt and solid, suggesting that there may be different temperature distributions in each phase. The uncertainty about the temperature distribution in a specimen in a concentric solidification experiment has hampered the overall understanding and further analysis of solidification phenomena, specifically modeling the morphology and rate of progression of the liquid/solid interface. Hence, the present study aims to determine experimentally the temperature distribution over the entire range of solidification in a specimen solidified in a concentric solidification configuration under dynamic cooling conditions.

Heat Transfer Mechanism

Experimental Setup

In spite of many attempts to determine the temperature distribution and solute profile in a concentric solidification experiment, particularly at high cooling rates, there are still not acceptable agreements between experiment and model predictions [15, 16] and there is a need to address the radiation heating mechanisms with reference to the existing literature [17]. The specimen is positioned at one focal point of a gold-plated ellipsoidal cavity within an infrared heating furnace, where a high integrity inert atmosphere is maintained. The heat source consists of a 1.0-kW halogen lamp located at the other focal point. A more detailed description has been provided elsewhere [2, 18]. The sample is heated by radiation, and the heat energy from the halogen lamp is partially distributed to the specimen. A B-type thermocouple is a spot welded onto the platinum specimen holder in order to provide a feedback signal. The power to the halogen lamp is controlled by an OMRON ES100P digital PID controller.

Heat Transfer Regions

By assuming that the temperature is symmetric around the center of the liquid pool, the temperature and heat distribution can be defined within two distinct regions (A, B) and one single point C of the 9.70-mm-diameter specimen disk.

In Region A, $0 \leq r_{\mathrm{m}} \leq 1$ mm (r_{m} = distance from the liquid center), there is a net flow of heat into the specimen (positive heat flow). At steady state, the radiation heat emitted from the bulb per a unit area E_{b}' is $E_{\mathrm{b}}' = \frac{P \times a\%}{A_{\mathrm{b}}}$. P is the power output of the bulb, and $a\%$ is the percentage of the power and is experimentally recorded by the PID controller. A_{b} is the area of the filament. However, only a fraction of the heat flux is received by the specimen and given by E_{b}. $E_{\mathrm{b}} = e_{\mathrm{f}} E_{\mathrm{b}}'$, where e_{f}' can be considered as a constant only in a limited temperature range (100–110 K).

In Region B, 1 mm < r_{m} < 4.85 mm, the net radiation heat input is negative. Heat escapes by radiation, convection (vertically), and conduction (radially).

At Point C, r_{m}=4.85 mm, the outer edge, heat is emitted in the vertical direction by the same mechanisms as in region B. At the outer edge of the sample, heat is emitted from the sample by convection and radiation alone. With these assumptions, a numerical model can be developed for transient heat transfer by the Gaussian iteration method [19], although model development forms part of the larger program, and such analysis is beyond the scope of the present study.

Experimental Methods

The Interface Temperature in the Concentric Solidification Process

The interface temperature depends upon both the radial position and time at a given cooling rate. During cooling, the specimen as a whole reduces it's temperature and the temperature gradient at the solid/liquid interface given by $\left(G_{\dot{T}}\right)$. The position of the interface in the radial direction toward the liquid center can be experimentally determined. For example, at a cooling rate of 10 K/min $r_m = (-0.0019 \times t) + 2.54$ for t: $0\ s \leq t \leq 230\ s$.

The interface temperature during concentric solidification under a given radial thermal gradient can be expressed as:

$$T_{\text{interface},t} = \left(T_{(5)-\text{surface},t} + K_{\dot{T}}\right) + \left((r_{\text{initial}} - r_m) \times G_{\dot{T}}\right) \quad (1)$$

r_{initial} is the initial liquid pool radius at $t = 0$ s, generally taken as $r_{\text{initial}} = 2.5$ mm. $T_{(5)-\text{surface},t}$ is the temperature recorded from a thermocouple attached to the outer edge of the sample (or on the Pt holder which is the same at slow cooling rates as shown in Fig. 5). The system is under equilibrium at $t = 0$ ($r_m = 2.5$ mm). $K_{\dot{T}}$ is calculated as $T_{\text{interface},t} = T_{eq}$, where T_{eq} is the equilibrium temperature for a liquid fraction at $t = 0$ ($r_m = 2.5$ mm). $K_{\dot{T}}$ is assumed to be a constant during the entire solidification time. Upon experimentally measuring $G_{\dot{T}}$, the interface temperature, $T_{\text{interface},t}$, can be calculated from $t = 0$ to $t = t_{\text{end}}$ (the entire solidification process).

Experimental Setup for Multiple Thermocouples in Solid State

For the purposes of measuring the temperature distribution in the solid state over the specimen radius, three of B-type thermocouples, 0.2 mm diameter, are spot welded onto a Fe–0.18mass%C steel disk with 0.23 mm thickness and 9.70 mm diameter. One thermocouple is welded at the geometrical center of the thin disk and the other two are at positions 2.0 mm and 3.0 mm radially away from the center as shown in Fig. 1. At the beginning of the experiment, the geometrical center of the sample is aligned with thermal center.

The following data is recorded with respect to time, for different cooling rates, varying from 2 K/min to 200 K/min from different starting temperatures:

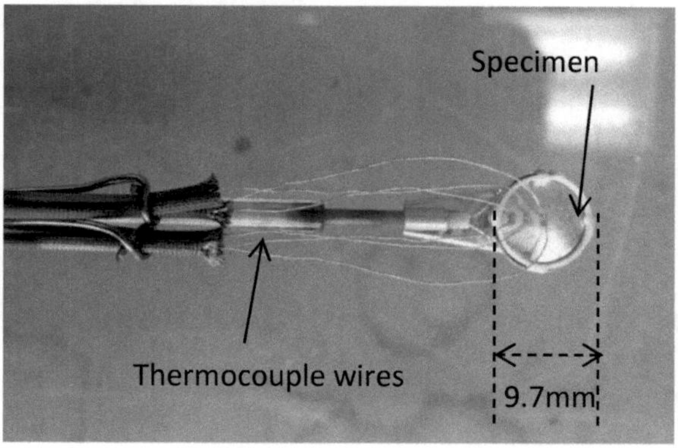

Fig. 1 Experiment setup of the specimen with three thermocouples spot welded

- Signal from four thermocouples (including the thermocouple attached to Pt-holder);
- Power input to the furnace;
- Video feed from the laser scanner at 30 fps [20].

Data is recorded at different cooling rates from different starting temperatures 1273, 1473, and 1573 K. The temperature gradient with respect to both cooling rate and initial temperature is further analyzed as input to the numerical model discussed in section "Heat Transfer Mechanism."

Experimental Setup for Multiple Thermocouples in the Solid–Liquid State

A vital part of the experiment is measuring the temperature in the liquid melt. The same experimental setup is used as mentioned in the section "The Interface Temperature in the Concentric Solidification Process" but with only two 0.2-mm-diameter thermocouple wires spot welded roughly 1–2 mm apart to the surface and then glued to the outer edge of the sample holder with alumina cement. The additional support from the alumina cement, constrained the wire to the sample holder, so that the thermocouples were suspended in the liquid pool surrounded by the solid rim as shown in the following Fig. 2. r_{Solid} and r_{Liquid} are the distances from the center to the thermocouples in solid and liquid, respectively.

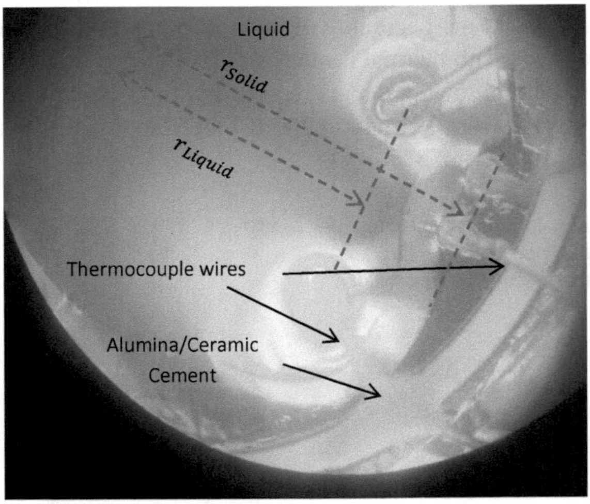

Fig. 2 B-type thermocouple immersed in the liquid pool

Accuracy of the Thermocouples and Calculations

The manufacture's accuracy ratings of the B-type thermocouples, Platinum–6%Rhodium and Platinum–30%Rhodium, respectively, are 1.5 K or 0.25% of reading, whichever is greater [21] [private communication]. At about 1770 K, the accuracy of the readings is within 4 K. During the experiment, the solidification process changed from a typical Fe–C solidification morphology when the thermocouples were immersed in the liquid for more than 30 min, suggesting that the thermocouples were partially dissolved into the liquid, thereby altering the initial thermocouple compositions and consequently the accuracy of the thermocouple readings. The distances from the liquid center are measured with image analyzing software, ImageJ [22]. The accuracy of the experimental measurements is further discussed in the following section.

Results and Discussion

Temperature Distribution in the Presence of a Solid–Liquid interface

Three separate experiments were conducted and two of them were conducted with only two thermocouples attached to the specimen surface with the solid/liquid interface located between them. Each experiment started by stabilizing a 5-mm-diameter liquid pool and the distance of the thermocouples are measured with respect to the center of the liquid pool. This methodology provides a means of mapping the thermal profile in the liquid as well as in the solid and importantly, the ability to calculate the temperature difference between the liquid and the solid across the interface.

In the first experiment, thermocouples T_1 and T_3 (Fig. 3) are located 1.91 mm and 2.82 mm away from the center of the liquid pool, respectively. In the second experiment, T_2 and T_4 are located 2.40 mm and 3.47 mm away from the center, respectively. In the third experiment, only one thermocouple is attached and located 3.60 mm away from the center.

Figure 3 shows an example of one set of such experiments, conducted at a cooling rate of 10 K/min. The dots and diamonds are experimental measurements and the solid curves represent a third polynomial fitting. The measured temperatures are shown with respect to the temperature node closet to the edge (T_5). Two curves are shown: One ($t = 0$ s) displays the temperature profile at the stable conditions of the experiment represented by the third polynomial, $T(r_{\mathrm{m}}) = -31.621r^3 + 242.24r^2 - 623.52r + 597.35$ and the other ($t = 60$ s) the temperature profile after 60 s at a cooling rate of 10 K/min.

Measurements of temperature are limited because of the solid/liquid interface progresses in the course of solidification. For example, in the experiment conducted at a cooling rate of 10 K/min, the solid/liquid interface reaches thermocouple T_2

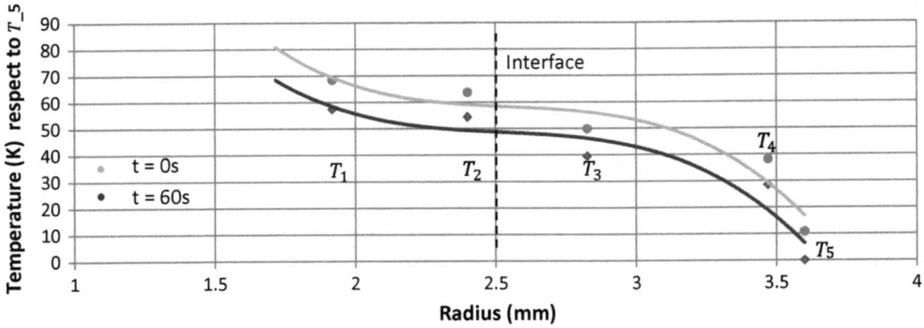

Fig. 3 Temperature distributions over the radial distance

Fig. 4 Thermal gradient versus cooling rate

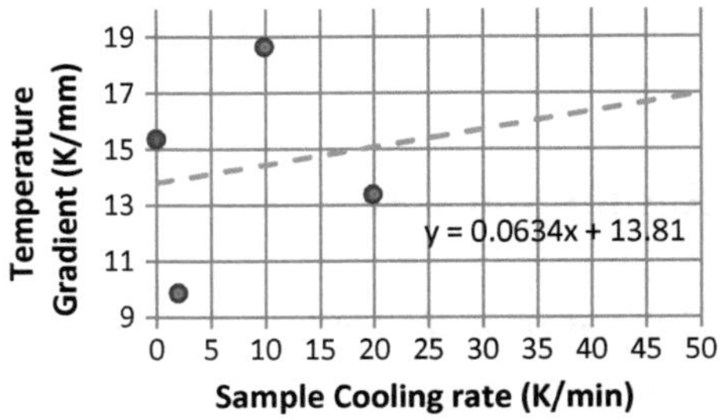

when it has progressed by 0.1 mm from its initial position (at a radius of 2.5 mm), and hence, experimental data can only be obtained for 60 s. In the experiments, temperatures are measured as a function of time and we averaged the linearized temperatures difference between T_1 and T_2, over time for a given solidification rate in order to determine the temperature gradient, in the liquid phase and these results are shown in Fig. 4. The temperature gradient increases slightly with an increase in cooling rate.

Turning now to Eq. (2):

$$T_{\text{interface},t} = \left(T_{(5)-\text{surface},t} + K_{\dot{T}}\right) + \left((r_{\text{initial}} - r_{\text{m}}) \times G_{\dot{T}}\right) \tag{2}$$

Since the thermal gradient has been measured experimentally, it is now possible to calculate the temperature at the solid/liquid interface at any given cooling rate.

[During the solidification of delta ferrite, the interface movement is less than 0.5 mm for at cooling rates higher than 10 K/min and less than 60 s in time. Hence, linearization of the temperature gradient for such small radial distances seems to be justified.]

Maintaining thermocouple wires immersed in the liquid found to be very difficult and liquid layer tends to rapture particularly during interface movement. Consequently, experimental data is limited to three different cooling rates, 2, 10, and 20 K/min and at isothermal holding. Temperature gradient variation respect to cooling rate is represented by the dashed line and given by the first-order polynomial,

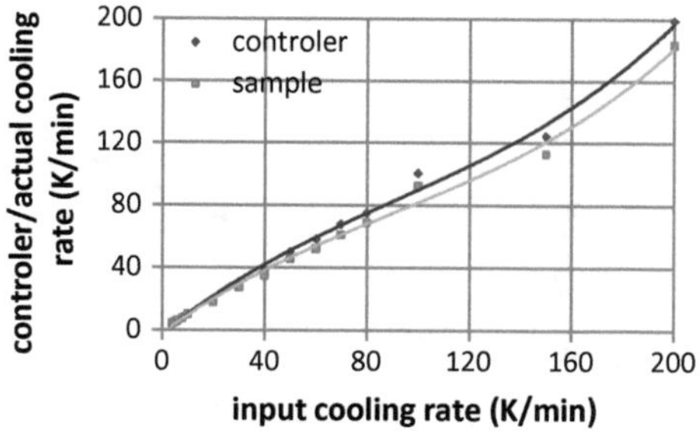

Fig. 5 Input cooling rate versus controller and sample cooling rate

$G_{\text{Temperature}} = \left(0.0634 \times \dot{T}\right) + 13.81$ in Fig. 4. The interface temperature can be calculated according to Eq. (2). Nodes T_1 and T_2 have specifically been chosen to determine the temperature gradient in the liquid since kinetic theory predicts [23], and it has been shown experimentally [24] that the nucleation and propagation of the reaction event occur on the liquid side of the solid/liquid interface.

Actual Cooling Rate on the Sample

In order to use Eq. (2), T_5, the cooling rate on the surface of the sample, is required. Although the temperature has been measured by attaching a thermocouple to the edge of the sample, this temperature in a normal concentric solidification experiment is taken as the control temperature, measured on the Pt holder (the control thermocouple). It is therefore important to calibrate the cooling rate measured by the control thermocouple against the actual cooling rate measured on the specimen surface, as shown in Fig. 5. At cooling rates lower than 10 K/min, the cooling rate indicated by the control thermocouple is almost identical to the true cooling rate measured on the edge of a specimen. However, at cooling rates higher than 40 K/min, the actual cooling rate on the sample is lower than that indicated by the control thermocouple as shown in Fig. 5. It is therefore important that this calibration curve is used in general concentric solidification experiments.

Simulation Results

By utilizing MICRESS, a two-dimensional multicomponent/multiphase commercial simulation package [25], the initial delta ferrite/liquid interface velocity for different cooling rates was calculated, starting with a 2.5-mm-diameter liquid pool for Fe–0.18mass%C specimen. Linearized thermodynamic data from ThermoCalc 2018b [26] was used in this calculation. The simulation domain is constructed as close as possible to the experimental setup and following the same methodology as

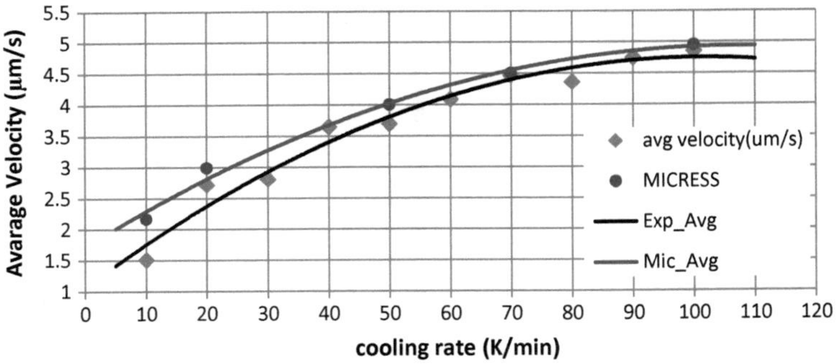

Fig. 6 Experimental (diamonds) and simulated (dots) average interface velocities at different cooling rates

earlier researchers [15, 27]. The experimentally determined values are compared to those calculated by MICRESS in Fig. 6.

The good agreement between the experimental and simulated solid/liquid interface velocities over a range of cooling rates provides confidence that the experimentally determined increase in thermal gradient in the liquid phase with cooling rate is reliable (Fig. 4).

Conclusion

1. Under isothermal conditions, the concentrically solidifying system is at steady state and the temperature profile with respect to radial distance can be expressed by a third polynomial as: $T(r_\text{m}) = -31.621r^3 + 242.24r^2 - 623.52r + 597.35$
2. At rapid cooling conditions, the temperature gradient in the liquid increases and is given by $G_\text{Temperature} = \left(0.0634 \times \dot{T}\right) + 13.81$
3. In a concentric solidification experiment, the input cooling rate and the actual cooling rate of the specimen varies. At low cooling rates (4–10 K/min), the decrease in the cooling rate on the specimen is less than 0.2 K/min, but at cooling rates as high as 100–200 K/min it varies from 8 K/min to 17 K/min, respectively.
4. The interface temperature for a concentric solidification experiment can be expressed as $T_{\text{interface},t} = \left(T_{(5)-\text{surface},t} + K_{\dot{T}}\right) + \left((r_\text{initial} - r_\text{m}) \times G_{\dot{T}}\right)$ and can be employed as an input parameter in a phase-field simulation.
5. The velocities of the solid/liquid interface measured experimentally and calculated by a phase-field simulation are in excellent agreement.

References

1. Hiroshi Chikama HS, Emi T, Suzuki M (1996) "In-situ" real time observation of planar to cellular and cellular to dendritic transition of crystals growing in Fe–C alloy melts. Mater Trans, JIM 37:620–626
2. Mark Reid DP, Dippenaar R (2004) Concentric solidification for high temperature laser scanning confocal microscopy. ISIJ Int 44
3. Shibata H, Arai Y, Suzuki M, Emi T (2000) Kinetics of peritectic reaction and transformation in Fe-C alloys. Metall Mater Trans B 31(5):981–991
4. Arai Y, Emi T, Fredfriksson H, Shibata H (2005) In-situ observed dynamics of peritectic solidification and transformation of Fe–3 to 5 at. Pct Ni alloys. Metall Mater Trans A
5. Kerr HW, Cisse J, Bolling G (1974) On equilibrium and non-equilibrium peritectic transformations. Acta Metall 22(6):677–686
6. Stefanescu DM (2006) Microstructure evolution during the solidification of steel. ISIJ Int 46(6):786–794
7. Griesser S (2013) In-situ study of the influence of alloying elements on the kinetics and mechanism of the peritectic phase transition in steel. Doctor of Philosophy thesis, Faculty of Engineering and Information Sciences, University of Wollongong
8. Trivedi R (1980) Theory of dendritic growth during the directional solidification of binary alloys. J Cryst Growth 49(2):219–232
9. Kurz W, Fisher DJ (1981) Dendrite growth at the limit of stability: tip radius and spacing. Acta Metall 29(1):11–20
10. Aaronson HI (2002) Mechanisms of the massive transformation. Metall Mater Trans A
11. Bower TF, Flemings MC (1967) Formation of the chill zone in ingot solidification. AIME MET SOC TRANS 239(2):216–219
12. Trivedi R, Shin JH (2005) Modelling of microstructure evolution in peritectic systems. Mater Sci Eng, A 413–414:288–295
13. Yin H, Emi T (2003) Marangoni flow at the gas/melt interface of steel. Metall Mater Trans B 34b
14. Phelan D (2002) In-situ studies of phase transformations in iron alloys. PhD thesis
15. Dominic Phelan MR, Dippenaar R (2006) Kinetics of the peritectic phase transformation: in-situ measurements and phase field modeling. Metall Mater Trans A
16. Griesser S et al (2014) Diffusional constrained crystal nucleation during peritectic phase transitions. Acta Mater 67:335–341
17. Griesser S, Reid M, Pierer R, Bernhard C, Dippenaar R (2014) In situ quantification of micro-segregation that occurs during the solidification of steel. Steel Res Int 85(8):1257–1265
18. Griesser S et al (2012) SolTrack: an automatic video processing software for in situ interface tracking. J Microsc 248(1):42–48
19. Holman J (2002) Heat transfer, 9th edn. McGraw-Hill
20. Marvin M (1961) Microscopy apparatus. Google Patents
21. Thermocoupleinfo. https://www.thermocoupleinfo.com/thermocouple-types.htm. Accessed 4 June 2018
22. Rasband W (2012) ImageJ software. National Institutes of Health, Bethesda, MD, USA, 1997
23. Schmelzer JWP, Röpke G, Priezzhev VB (1999) Nucleation theory and application
24. Griesser S, Bernhard C, Dippenaar R (2014) Mechanism of the peritectic phase transition in Fe–C and Fe–Ni alloys under conditions close to chemical and thermal equilibrium. ISIJ Int 54(2):466–473
25. Steinbach I, Apel M (2013) MICRESS the MICRostructure evolution simulation software
26. Andersson J-O et al (2002) Thermo-calc & DICTRA, computational tools for materials science. Calphad 26(2):273–312
27. Tiaden J (1999) Phase field simulations of the peritectic solidification of Fe–C. J Cryst Growth 198/199:1275–1280

In Situ Investigation of Pt–Rh Thermocouple Degradation by P-Bearing Gases

Anna Nakano, Jinichiro Nakano and James Bennett

Abstract Gases bearing elements such as As, S, and Si are known to degrade Pt and Pt–Rh alloys, leading to thermocouple sensor failure during industrial operations. While the corrosive impact of As, S, and Si gases have been discussed in the literature, the impact of P has not been well studied. P may originate from the carbon feedstock, ores, additives, and refractory bricks used in metallurgical and gasification processes. In this work, gaseous P interactions with Pt–Rh (Rh = 6 and 30 wt%) alloys were isothermally investigated at 1012° C in situ using a customized environmental white light/confocal scanning laser microscope. Changes in microstructure and P-diffusion into the Pt–Rh alloys are discussed based on real-time images recorded during the exposure tests and electron probe microscopy analysis from the quenched samples.

Keywords Platinum · Rhodium · Phosphorous · Sensor failure · Corrosion Thermocouples

Introduction

Pt–Rh thermocouples are critical to many applications ranging from laboratory testing to industrial processing, but the loss of the sensor would impact those practices. Sensor failure can be caused by a number of reasons related to service environments, including chemical attack, abrasive wear, and shear. As, S, and Si are known to degrade Pt and Pt–Rh alloys, leading to thermocouple sensor failure during industrial operations. However, the impact of P is not well understood or reported.

In earlier research by the present authors, failure analysis of the wires (Pt–30 wt%Rh and Pt–6 wt%Rh) from a type B thermocouple and Pt–Rh alloys with Rh

A. Nakano (✉) · J. Nakano · J. Bennett
U.S. Department of Energy National Energy Technology Laboratory,
1450 Queen Ave, Albany, OR 97321, USA
e-mail: anna.nakano@netl.doe.gov

A. Nakano · J. Nakano
AECOM, P.O. Box 618, South Park, PA 15129, USA

J. Nakano et al. (eds.), *Advanced Real Time Imaging II*, The Minerals, Metals & Materials Series, https://doi.org/10.1007/978-3-030-06143-2_14

varied from 0 to 30 wt% was conducted by non-isothermal [1] and isothermal [2] techniques, respectively, in P-rich gas environments using a high-temperature resistance furnace. Two distinct failure mechanisms driven by the P-diffusion into the Pt–Rh alloys exhibited a correlation with Rh content: (1) intergranular diffusion in Pt–6 wt%Rh and (2) intragranular diffusion in Pt–30 wt%Rh. The failure in the 30 wt%Rh alloy occurred as grains were corroded through interactions with incoming P, while that in the 6 wt%Rh alloy initiated along grain boundaries that liquefied due to the P-enrichment. The failure of type B thermocouple wires was found to occur at temperatures as low as 929–979 °C. In support to the previous study based on post-analysis for the Pt–Rh alloy failure mechanisms, real-time observations in the current work revealed initial and temporal stages of microstructural changes as P interacted with the alloys.

Experimental

Materials

The thermocouple grade Pt–6 wt%Rh and Pt–30 wt%Rh wires of 0.38 mm diameter manufactured by Engelhard were surface polished to 1 micron. Po–S NONDRY, Struers ApS, containing silica and water was used as a polishing agent. The 99.6% purity aluminum phosphate powder, <5 μm in diameter, supplied by Alfa Aesar was used as the source of gaseous P species. The X-ray diffraction (XRD) analysis of the aluminum phosphate powder indicated three crystalline phases: 92.4 wt%$AlPO_4$, 7.2 wt%AlP_3O_9, and 0.4 wt%Al_2O_3. All the gases used in the tests were high purity and supplied by Matheson: 99.999%N_2, 99.5%CO, 99.999%He and a 23 vol%O_2–N_2 gas mixture.

Test Procedures

Gaseous P interactions with Pt–Rh (Rh = 6 and 30 wt%) alloys were investigated at 1012 °C isothermally in situ using a customized environmental white light/confocal scanning laser microscope (CSLM, Olympus-OLS3100 with Yonekura-15TPS). Detailed descriptions of the CSLM heating chamber used can be found elsewhere [3]. A schematic diagram of the sample setup is shown in Fig. 1. About 0.017 g of the aluminum phosphate powder was placed in a high-density alumina crucible. The sample wire was positioned on the crucible rim above the powder with the polished side facing up. Two alumina lids were placed on each side of the wire to keep the wire in place during the test and to minimize a P-gas loss into the furnace chamber. Purge gases were introduced onto the sample directly from above.

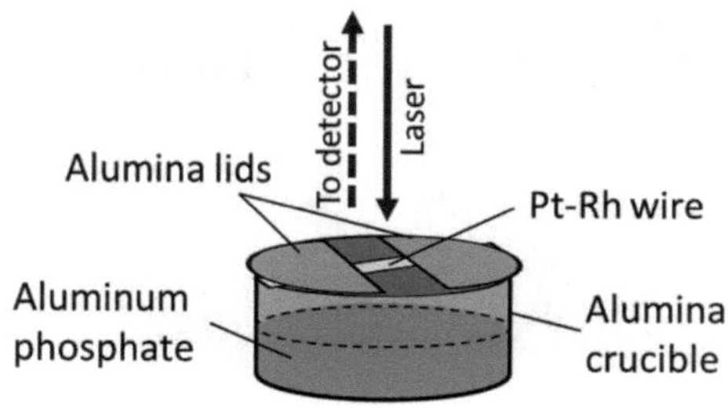

Fig. 1 A schematic diagram of the sample setup

Changes in microstructure of the Pt–Rh alloy were observed in situ by heating a test material to 1012 °C at a rate of 1000 °C/min in 5 vol%O_2–N_2 flow of 50 mL/min. The 5 vol%O_2–N_2 environment was chosen to prevent aluminum phosphate thermal decomposition during heating and was created by mixing an appropriate ratio of N_2 and a premixed 23 vol%O_2–N_2 gas using Alicat mass flow controllers. After ensuring the temperature was stabilized (held for 5 min), the purge gas was switched to CO at a 50 mL/min flow rate to promote aluminum phosphate decomposition and the release of P-gases. After a 15-min exposure in the P-rich CO environment, the sample was quenched to room temperature by turning off the furnace and blowing a He gas at 60 mL/min. Real-time microstructural changes were recorded throughout the test with a resolution of 10 frames per second. The reference samples were produced using the same test procedure but without the presence of aluminum phosphate powder. The surfaces of the quenched samples were analyzed using electron probe microscopy (EPMA, JEOL JXA 8530F Plus Hyper Probe).

Results and Discussion

In Situ Microscopic Study of the Pt–Rh Alloys

CSLM laser images (non-confocal mode = less sensitive to a specific wavelength) of the reference samples and those exposed to the P-rich CO environment are shown in Figs. 2 and 3, respectively.

In the reference sample alloys, grain boundary opening was noted during heating and the 5 min hold at 1012 °C in 5 vol%O_2–N_2. Also, small pores became visually noticeable. Grain boundaries continued to widen after gas switching to CO. No other apparent changes were observed on the alloy surfaces during the reference tests. Note that, in general, grain sizes seemed to vary drastically between both alloys ranging from less than 3 μm to about 40 μm in the Pt–30 wt%Rh alloy, and from about 10 μm to more than 100 μm in the Pt–6 wt% alloy.

In contrast to the reference samples, shortly after the purge gas was introduced, those exposed to the P-rich CO environment showed dramatic changes in microstructure (in addition to grain boundary opening) (see Fig. 3). According

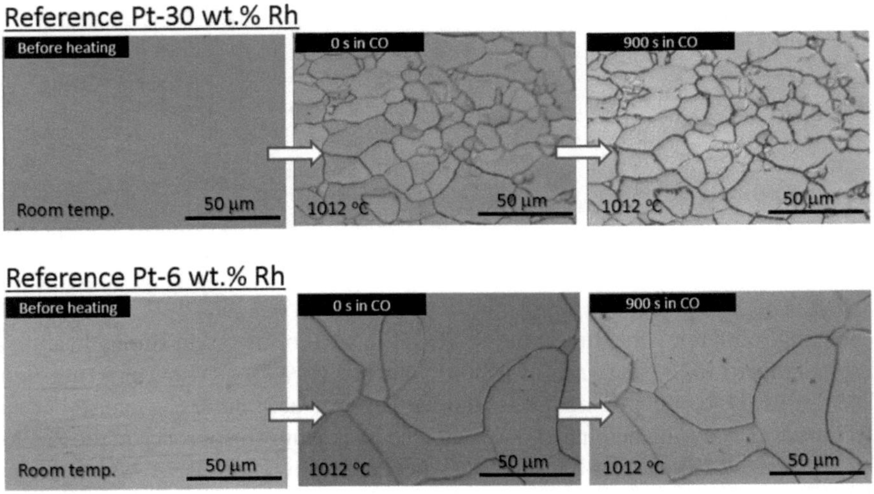

Fig. 2 CSLM laser images of the reference Pt–Rh alloys exposed to CO environment at 1012 °C without the presence of aluminum phosphate

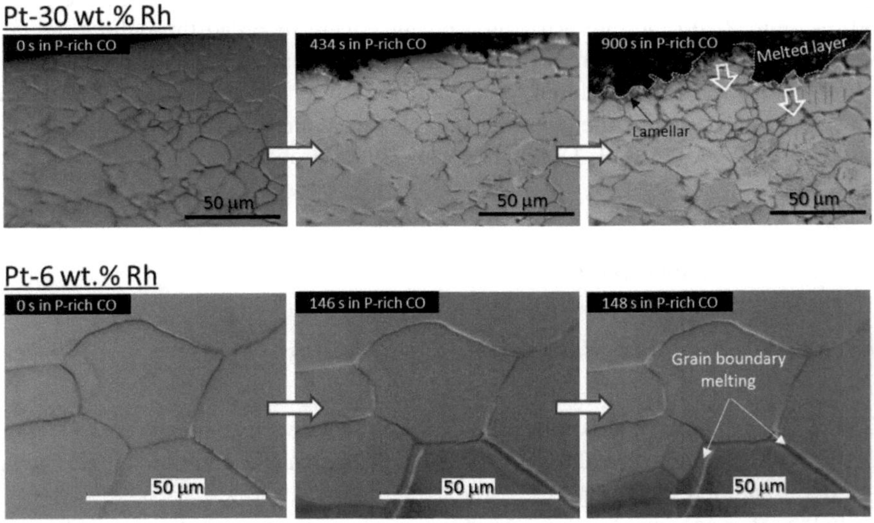

Fig. 3 CSLM laser images of the Pt–Rh alloys exposed to P-rich CO environment at 1012 °C with the presence of aluminum phosphate

to the thermodynamic calculations using FactSage 7.0 software with FactPS and FToxid databases [4], aluminum phosphate under the current test conditions in the presence of CO gas may decompose into various P-bearing gases with $(P_2O_3)_2$ being dominant. These P-bearing gas species are expected to interact with Pt–Rh wires, acting as the primary sources of P into the alloys. According to the binary P–Pt [5] and P–Rh [6] partial phase diagrams available in the literature, as P diffuses into the alloy, the Pt liquidus will be lowered and the formation of intermediate P–Pt and P–Rh phases are possible at the experimental conditions. Note, due to the sample configuration, higher thermodynamic activity of P in the gas phase was expected near the aluminum phosphate powders, which promoted early reactions of P-bearing gases with the non-polished surface (bottom) and sides of the wire (both locations are not visible from the confocal microscope), than with the polished side (top surface).

For the Pt–30 wt%Rh alloy, 120 s. after the purge gas was switched from the 5 vol%O_2–N_2 gas mixture to CO, dark pits (less than 1 μm in diameter when noticed) started to appear on the grain surfaces. The pits then grew in a uniform direction, which is peculiar to each grain (likely along (111) planes), becoming elongated microstructures. EPMA analysis indicated those dark microstructures were rich in Rh and P but was not conclusive due to their limited sizes. They may be related to the initial formation of the $(Pt, Rh)_2P$ phase, which was previously found in the lamellar structure from the cross-sectioned alloys [1]. The dark platelets forming the lamellar structure are expected to originate from grain boundaries. As P diffuses substantially faster through grain boundaries than through grains, $(Pt, Rh)_2P$ would form at the supersaturated grain boundaries and grow into the grains as plates. P may enter grain boundaries from the surface or melted zones, which would affect the growth direction of the platelet tips on the surface. If P enters from the melted zone, it is possible $(Pt, Rh)_2P$ forms at the liquid/solid interface or grain boundaries underneath and grow upward, other than on the surface and grow downward. This may explain the surface pit formation and growth within the grains. In about 200 s. in the P-rich CO gas, the edge of the wire (top in figure) began to melt. The melted layer grew into the center of the polished surface, implying the governing direction of P diffusion was from the edge to the center (top to center in the figure). The platelets were noted growing ahead of the melting layer. This will be further discussed in the next section.

For the Pt–6 wt%Rh alloy, 109 s. After the CO gas was introduced into the furnace chamber, liquid started to form along the grain boundaries. This is in agreement with the previous work [1, 2] where grain boundary opening followed by the formation of the intermediate $(Pt, Rh)_2P$ phase was the key degradation mechanism for this alloy composition based on cross-sectional analysis of the quench-in experiments. The formation of the $(Pt, Rh)_2P$ phase was not clear from the CSLM real-time images alone. However, EPMA post-analysis of the recovered Pt–6 wt%Rh alloy indicated the presence of the $(Pt, Rh)_2P$ phase in the Rh-poor Pt liquid enriched with P. An example of the grain boundary melting during the exposure test is shown as temporal successive images in Fig. 3. Grain boundaries started to melt, expanding into neighboring grains. Eventually, liquid covered the entire wire surface. By 570 s, most of the sample was severely melted under the experimental conditions.

Phase Transformations of the Pt–30 Wt%Rh Alloy upon Isothermal P Exposure

In previous research by the authors [1, 2], degradation mechanisms of the Pt–Rh alloys depending on the Rh concentrations were proposed based on post-analysis of non-isothermal and isothermal quench-in tests. In the exposed Pt–30 wt%Rh wires, two distinct layers with a transition zone were present above the original alloy matrix. The outer layer contained blocky $(Pt, Rh)_2P$ particles in a Pt-rich phase (assumed to be liquid at the experimental temperature). The inner layer exhibited the lamellar structure growing into the Pt–Rh grains. It was suggested that the lamellar structure formed first, followed by the liquid layer. The present real-time observations confirmed such a sequence of phase transformations. The SEM image (Fig. 4) of the reacted surface shows an example of the advancing outer liquid layer with platelets growing ahead of it. EPMA analysis showed an atomic ratio of $(Pt+Rh){:}P \approx 2{:}1$, indicating the plates are likely the $(Pt, Rh)_2P$ phase.

Note, the EPMA analysis of the reference samples as well as the exposed samples indicated the presence of silicon on the polished surface and slightly higher concentrations in the grain boundaries. The silicon likely existed as silica introduced from the polishing media. The silicon concentration was relatively uniform across the sample surfaces with no variations between different phases, implying the presence of a thin silica film. The silicon concentration was difficult to obtain due to its thickness. The effect of the silicon is not certain, but it did not affect the degradation mechanisms in both alloys under the present experimental conditions. Further investigations are underway.

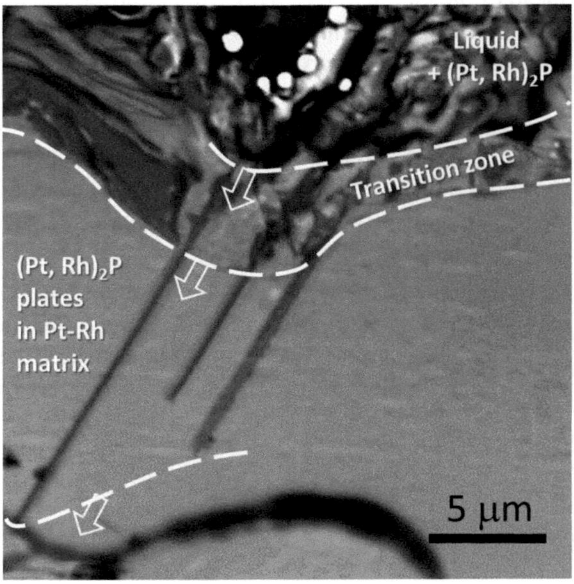

Fig. 4 SEM backscatter image of the quenched Pt–30 wt%Rh alloy that was exposed for 15 min to the P-rich CO environment at 1012 °C

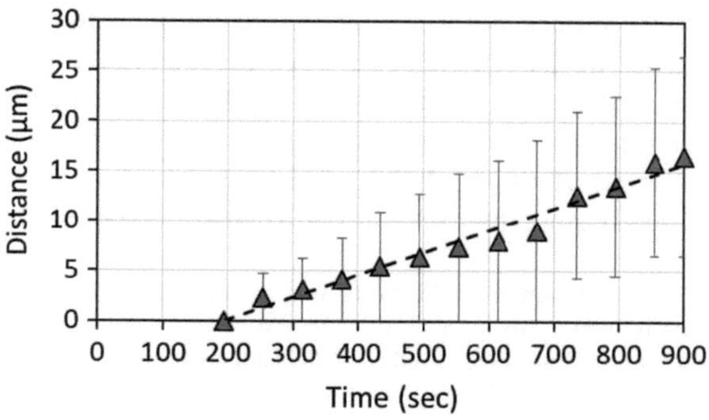

Fig. 5 Kinetics of the liquid layer growth on the surface of the Pt–30 wt%Rh alloy exposed to the P-rich CO environment at 1012 °C

Kinetics of the liquid layer growth over the surface of the Pt–30 wt%Rh alloy was analyzed from the CSLM movie. Results are presented in Fig. 5 as a change in layer width measured from the wire edge to the frontline. The location of the front line was assumed to be at the midpoint between the two parallel lines drawn at the extremes of the interface. The distances from the midpoint to the parallel lines were assumed as an uncertainty of the measurements. Time '0' on the plot indicates the moment of the CO gas flow was initiated.

It was found from Fig. 5, the liquid layer advanced over the polished surface at a rate of 0.02 μm/s. The linearity of the regression implies constant growth with steady-state interdiffusion of Rh and P across the layers. The diffusion field in Pt–Rh alloy can be modeled with a steady-state solution:

$$\gamma_{\mathrm{P}}(x,t) = \gamma_{\mathrm{P}}(X_1,t)\exp\left[-\frac{v(x-X_1)}{D_{\mathrm{P}}^{\gamma}}\right] \tag{1}$$

where γ_{P} is the concentration of P, x is the distance from free surface, t is the time, X_1 is the boundary location between the inner (lamellar) and outer (liquid) layers, v is the velocity, and D_{P}^{γ} is the diffusivity of phosphorus in Pt–Rh alloy. In the present work, the velocity of the liquid layer advancement was found to be approximately six times slower than that determined in the previous work. It is not certain if this was caused by the presence of silicon (as mentioned earlier—thought to originate from the polishing media), directions of P-diffusion, or experimental configurations. Further kinetics analysis is beyond the scope of this manuscript and will be discussed elsewhere.

Conclusions

Corrosion of Pt–Rh thermocouple alloys during exposure to P-bearing gases was investigated in situ using a customized environmental white light/confocal scanning laser microscope at 1012° C. Real-time observations of the P interaction with the

Pt–Rh surface confirmed the failure mechanisms proposed by the present authors in earlier works. The lamellar structure was developed and grew ahead of the liquid layer advancing into the Pt–30 wt%Rh alloy. Grain boundary melting was noted as the initiation of the material degradation in the Pt–6 wt%Rh alloy. A constant liquid layer growth in the Pt–30 wt%Rh alloy was noted, which was 0.02 μm/s, implying a steady-state interdiffusion of Rh and P across the layers.

Acknowledgements This work was performed in support of the U.S. Department of Energy's Fossil Energy Advanced Gasification Program. The research was executed through NETL Research and Innovation Center's Advanced Gasification effort. Research performed by AECOM Staff was conducted under the RES contract DEFE0004000. The authors acknowledge Prof. J.E. Morral (The Ohio State University) for valuable discussions, Mr. K. Collins (NETL) for SEM-EPMA, and M. Fortner (NETL) for metallography.

References

1. Nakano A, Bennett J, Nakano J (2016) Failure mechanisms in Pt–Rh_x thermocouple sensors caused bygaseous phosphorous species. Corros Sci 103:30–41
2. Nakano A, Nakano J, Bennett J (2017) Mechanisms of Pt–Rh thermocouple failure by gaseous phosphorous in high temperature processes. Paper presented at the UNITECR 2017, Santiago, Chile, 26–29 September 2017
3. Nealley WHH, Nakano A, Nakano J, Bennett JP (2018) Structural changes and material transport in Al_2O_3-supported $CuFe_2O_4$ particles in a simulated chemical looping combustion environment. JOM 70(7):1232–1238
4. FactSage chemical thermodynamic software, http://factsage.com/
5. Okamoto H (1990) P–Pt (phosphorus–platinum) system. ASM International Bulletin. Alloy Phase Diagrams 11:511–513
6. Okamoto H (1990) P–Rh (phosphorus–rhodium) system. ASM International Bulletin. Alloy Phase Diagrams 11:411–417

Author Index

J. Nakano et al. (eds.), *Advanced Real Time Imaging II*, The Minerals, Metals & Materials Series, https://doi.org/10.1007/978-3-030-06143-2

Subject Index

J. Nakano et al. (eds.), *Advanced Real Time Imaging II*, The Minerals, Metals & Materials Series, https://doi.org/10.1007/978-3-030-06143-2